SCULPTING TRADITIONAL BOWLS

Rip & Tammi Mann

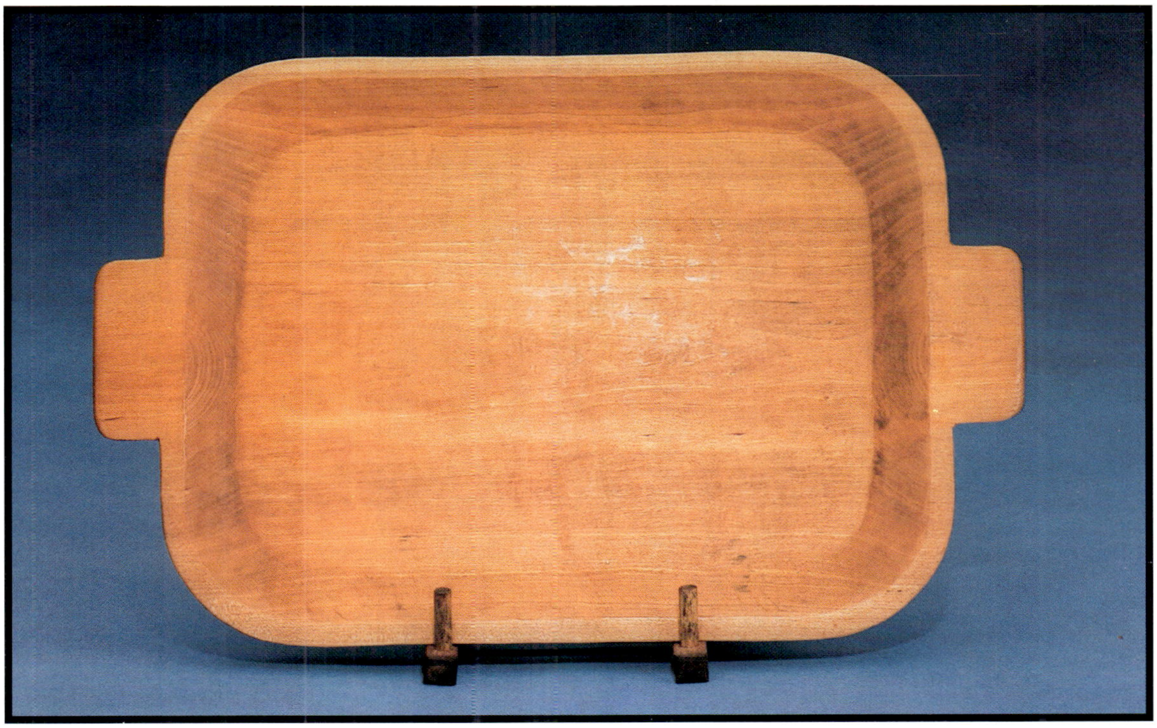

Text written with & photography by Douglas Congdon-Martin

Schiffer Publishing Ltd
77 Lower Valley Road, Atglen, PA 19310

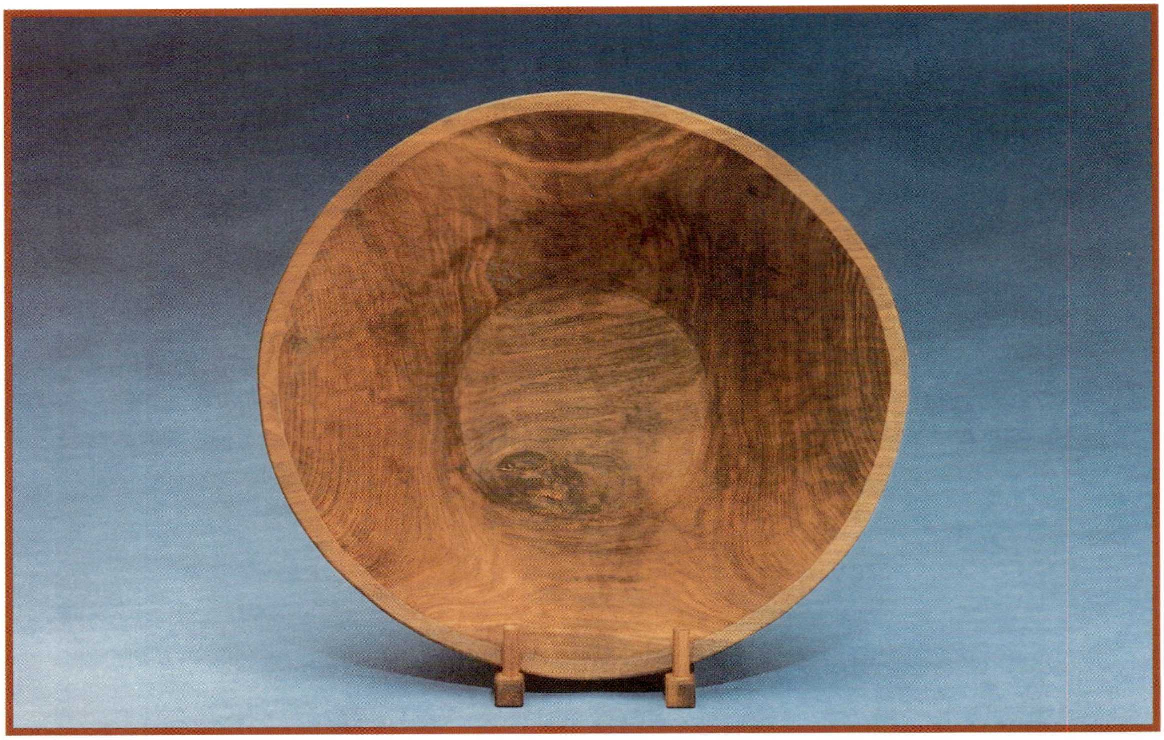

Contents

Introduction	3
Hewing the Bowl	4
The Seasoning Process	37
The Gallery	39

Printed in China
ISBN: 0-88740-698-X
We are interested in hearing from authors with book ideas on related topics.

Published by Schiffer Publishing Ltd.
77 Lower Valley Road
Atglen, PA 19310
Please write for a free catalog.
This book may be purchased from the publisher.
Please include $2.95 postage.
Try your bookstore first.

Introduction

This book is for the person who wants the personal joy and satisfaction of creating a useful item and a thing of beauty with one's own hands and heart. Remember, you leave a part of yourself in everything you "touch" (create) with your hands.

This is a return to the "good old days" and the old ways, using hand tools and a living piece of wood to make a bowl.

From the beginning, hewn wooden bowls have been very special to people. Anthropologists tell us that the "cave man" used to spend his winter months in his cave, with a stone adze and a chunk of wood, hewing away to create a bowl -- a useful item to be sure, but also a creation of beauty.

Wooden bowls and adzes accompanied people as they crossed the land bridge from Asia to North America. They were among the few items treasured enough to be carried along this journey of many lifetimes.

Until the advent of mechanization, a man would still spend his winter months in his home, using his metal adze to hew wooden bowls and treenware ("from a tree"). Each bowl so created has a part of its creator in it, distinguishing it from the latter-day, machine-made, production-line bowls.

One of the oldest crafts, hewing wooden bowls is *almost* a "lost art" in our modern land. Most people don't have the luxury of time to spend learning how to hew bowls, thus they turn to the mechanical methods of bowl making.

You can have the simple pleasures of creating a bowl your family can use and treasure for years and lifetimes to come. We are pleased to share with you the traditional method of Appalachian-style bowl hewing. This style of bowl originated with the Scots many centuries ago, and they brought it with them when they moved to America and settled in the Appalachian Mountains.

Although the modern world doesn't generally make bowls in this manner, travellers will have seen it practiced in many third-world countries until recently. As the world becomes more mechanized, more and more of these ancient cultures and ways are being lost.

Come along, be an adventurer, and return with us to a simpler time...a time when a person created with hands and heart, providing for loved one's needs by doing, not buying.

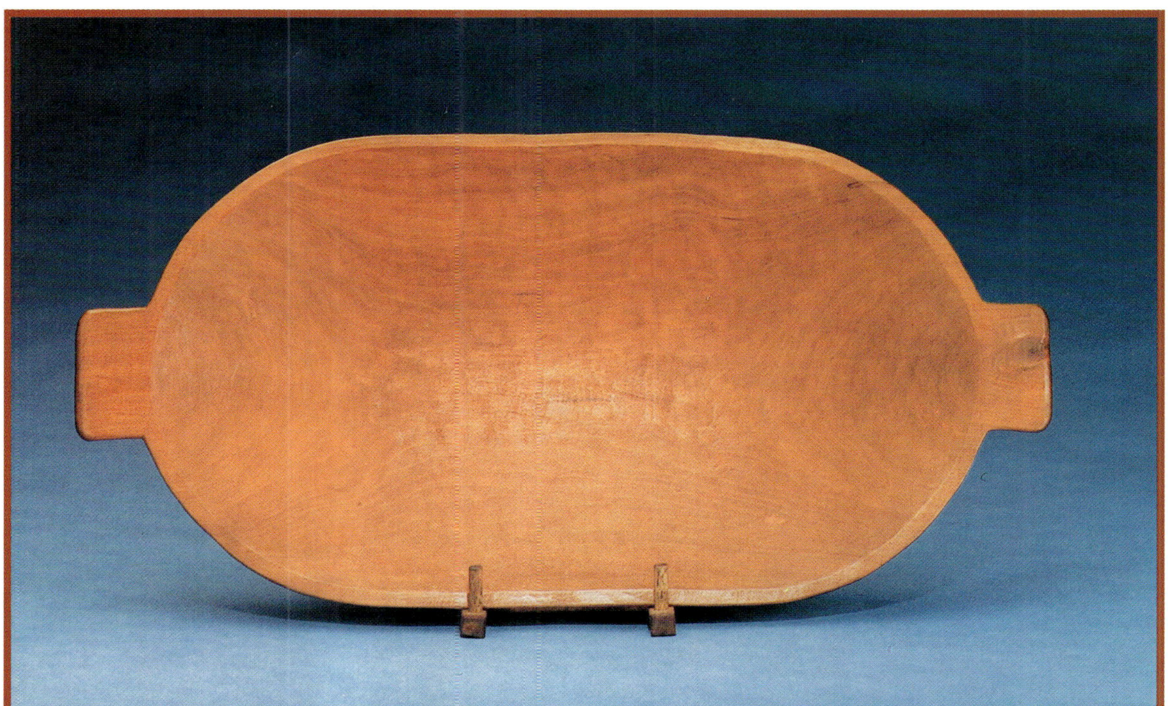

Hewing the Bowl

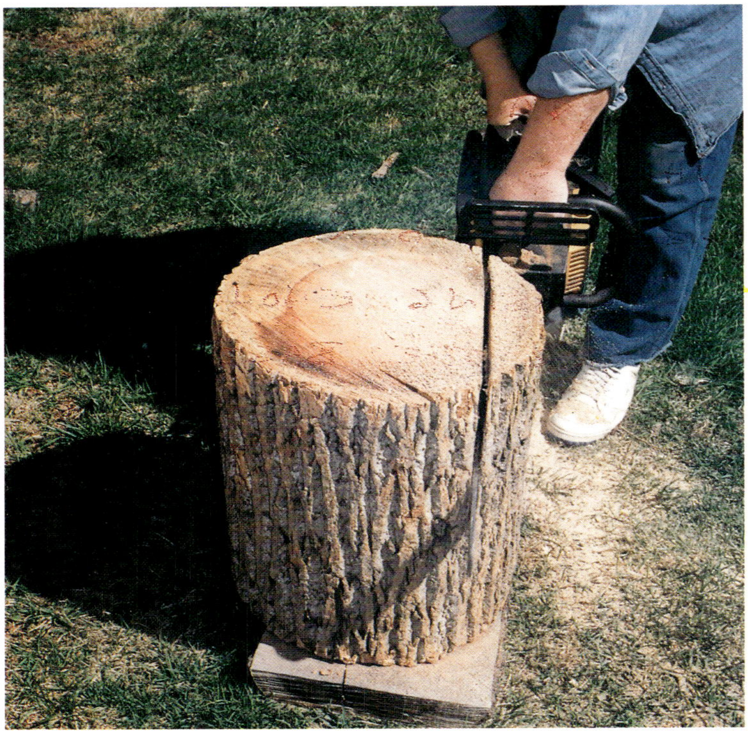

Bowl slabs are cut from a section of a tree approximately 4" longer than the finished length of the bowl. The slabs are cut with the grain of the wood, *not* cross-grain.

Begin by taking off an outside slab. This should be deep enough into the log to remove the sapwood, exposing enough heart-wood to allow for the base of the bowl. Cut as straight as possible.

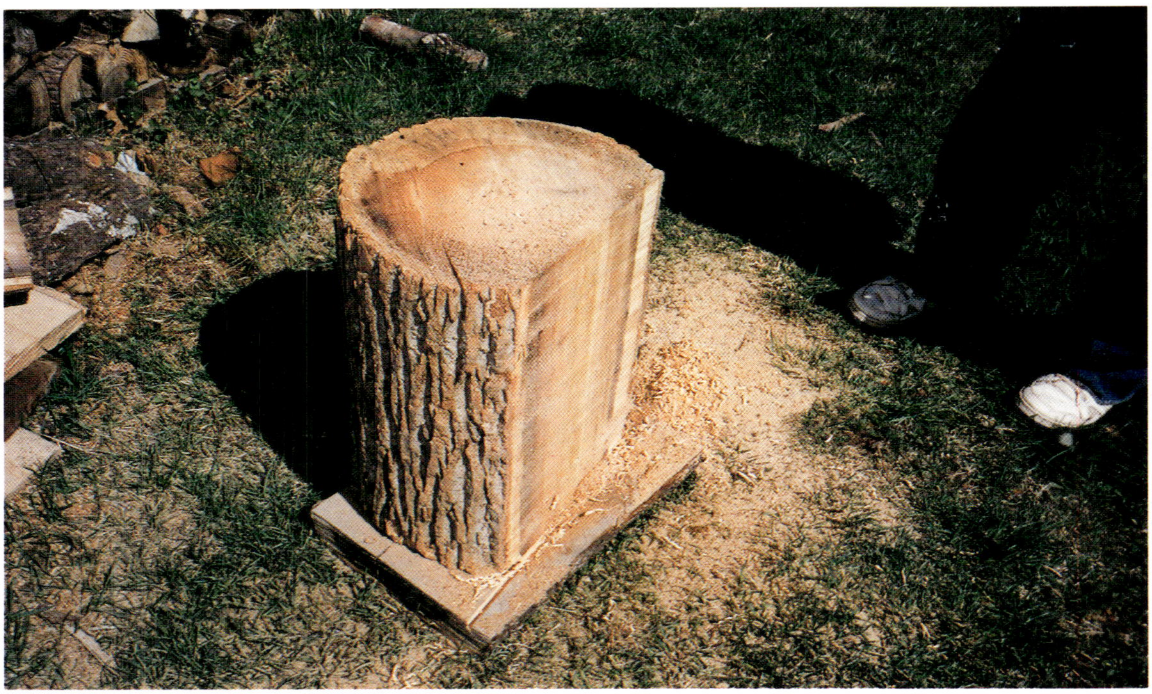

You can see that we have removed enough sapwood, leaving a good heart-wood base.

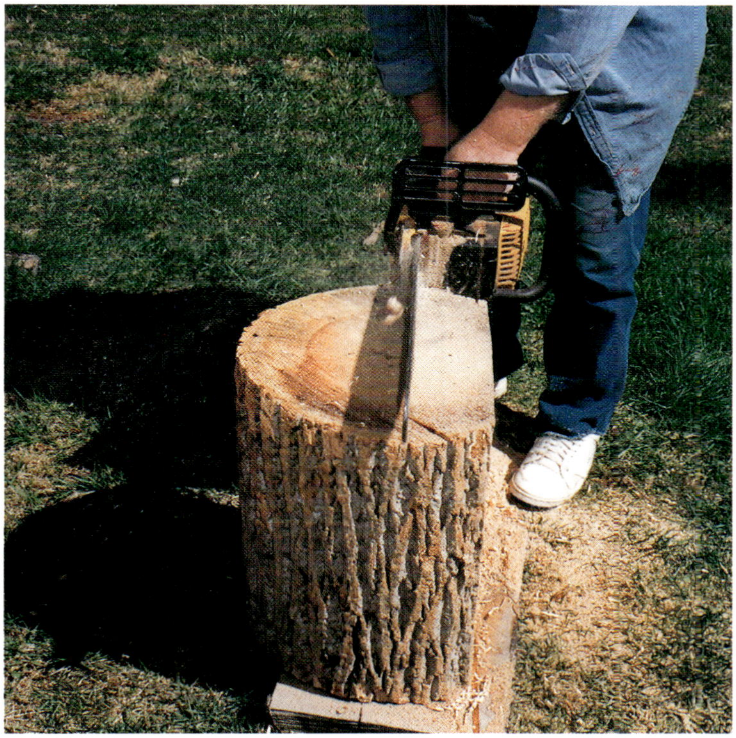

The second cut should be parallel to the first. The thickness of the slab will determine the depth of the bowl, remembering that the surfaces need to be planed smooth. Always stay 1 1/2" to 2" back from the pith (center growth rings).

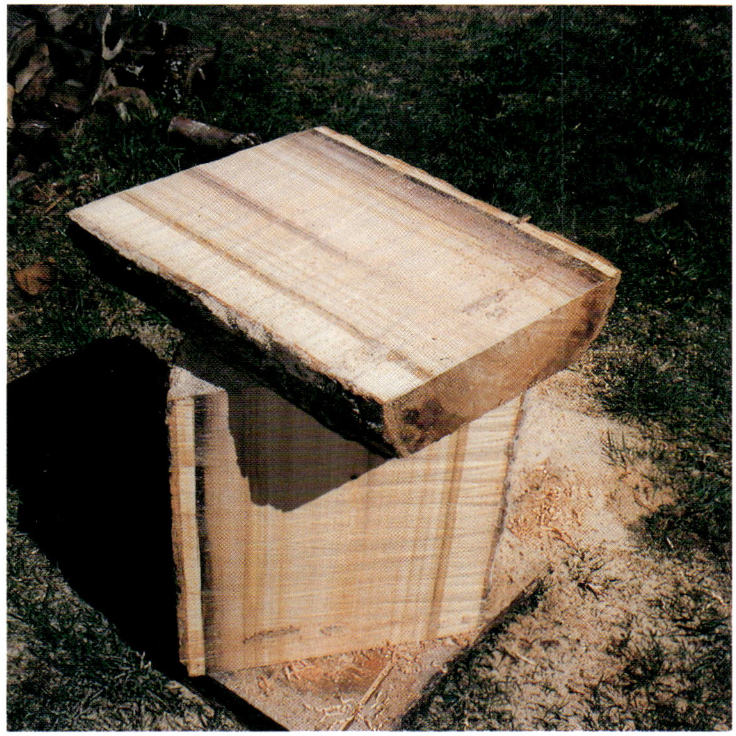

The slab will look like this. For hewing I use green wood because it works much easier with the adze. The finishing process we use keeps the bowl from splitting or warping.

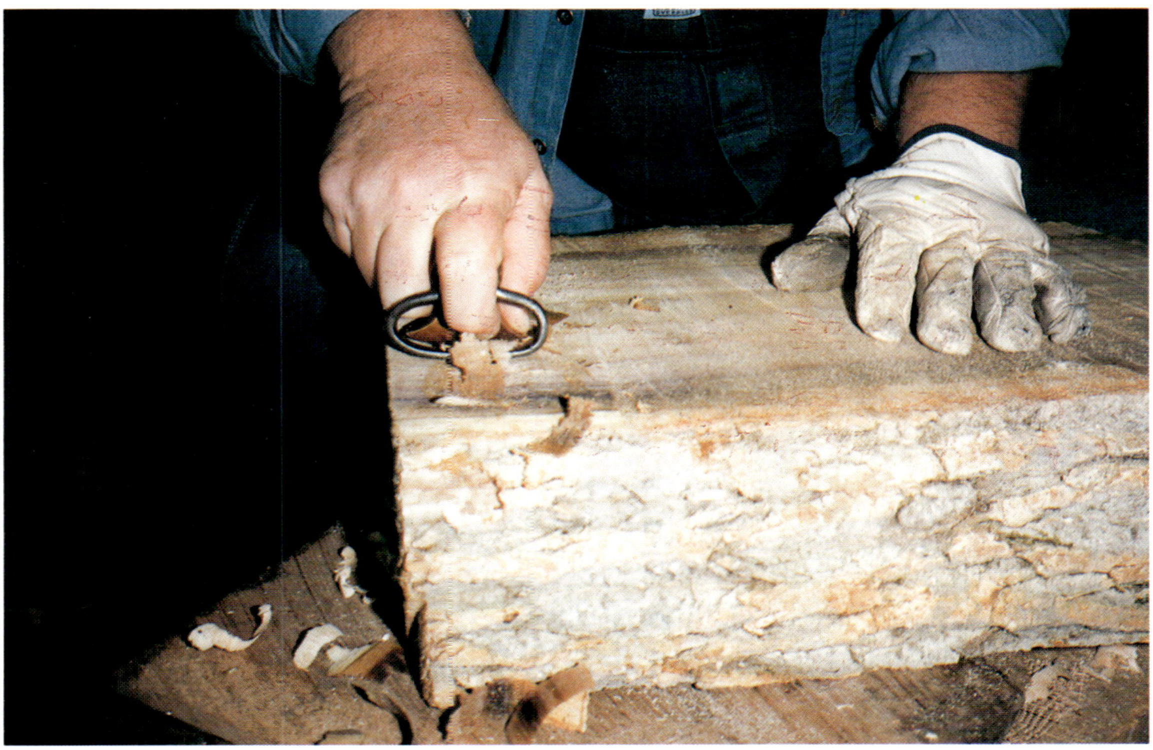

Plane both surfaces as smoothly and as parallel as possible. Once this is complete we are ready to lay out our bowl.

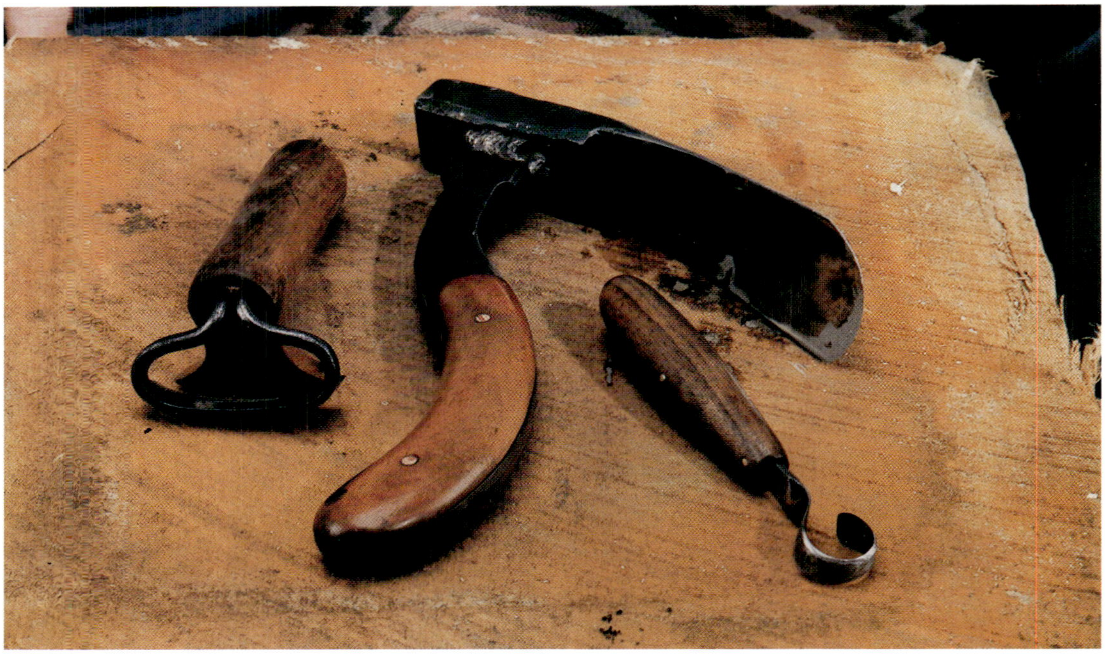

The tools for hewing bowls are simple. From left to right these are a bowl shave, a hand adze, and a scorp.

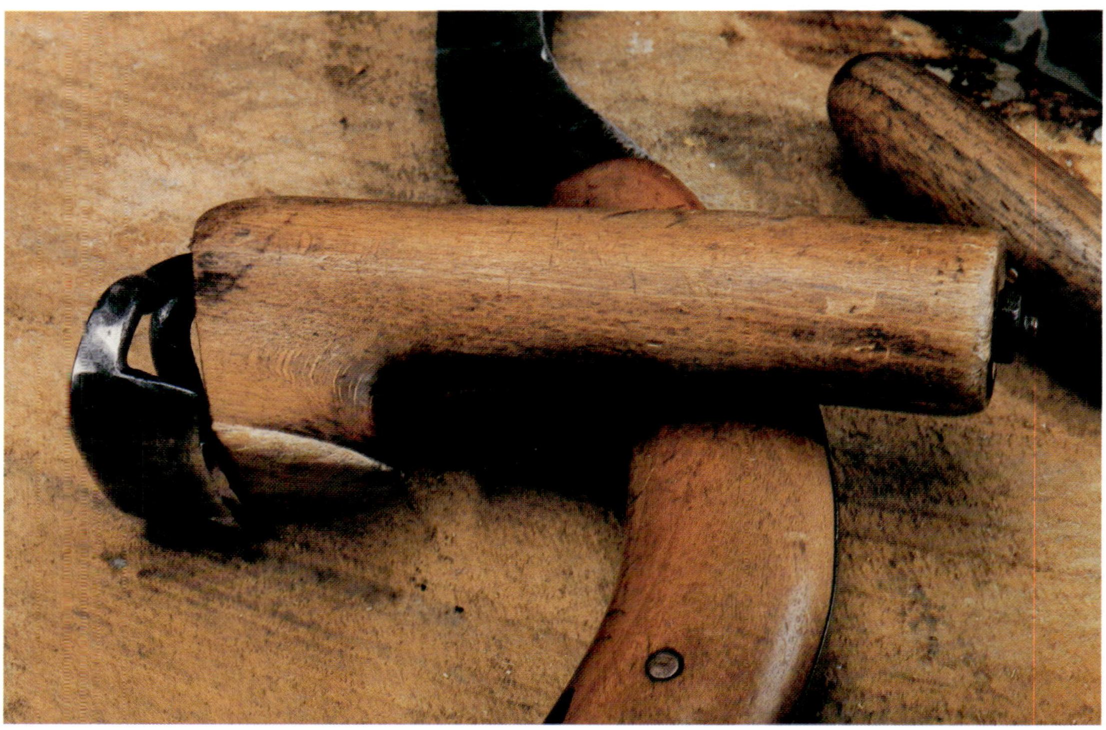

This bowl shave is 200 years old. It works like a plane. You may be able to find this from some antique tool dealer, as I did. If not you could have one made by a talented blacksmith, use a wood rasp like a Shur-form*TM or a small plane.

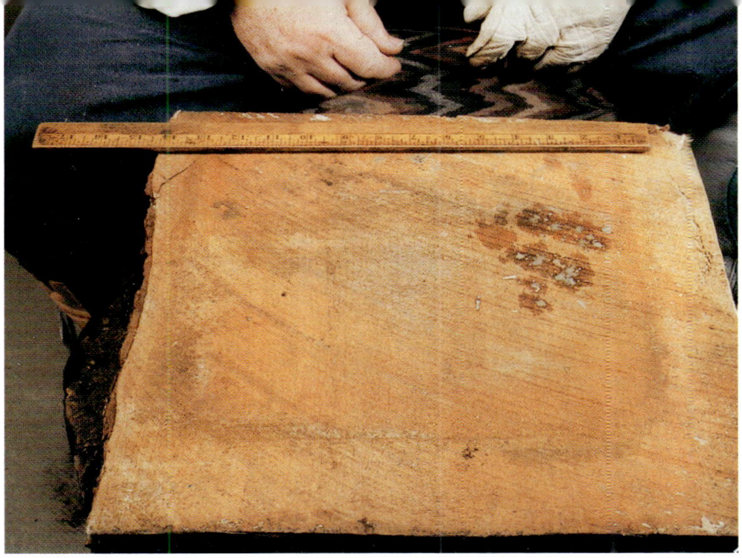

I'm making a rectangular bowl, 11' wide by 16" long, using a rectangular piece of cherry, three inches thick.

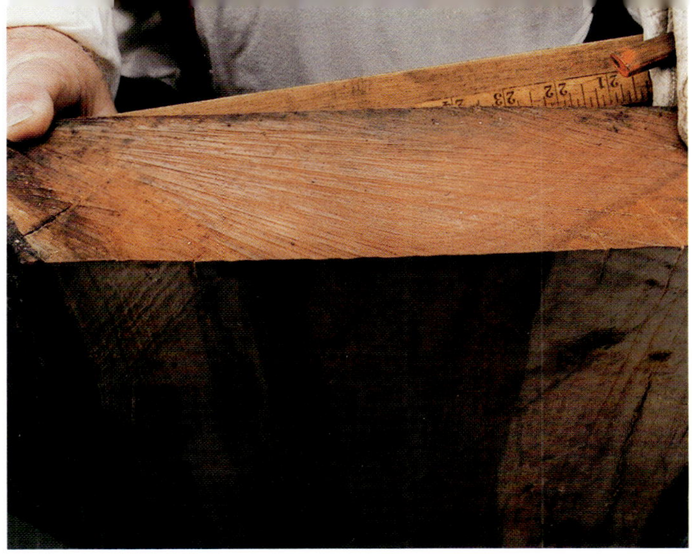

The color transition indicates the sap wood at the edge of the slab. I try to use as little as possible of this because it dries at a different rate than the heart wood does. If you use much of it, it is possible that your bowl could twist or warp a little.

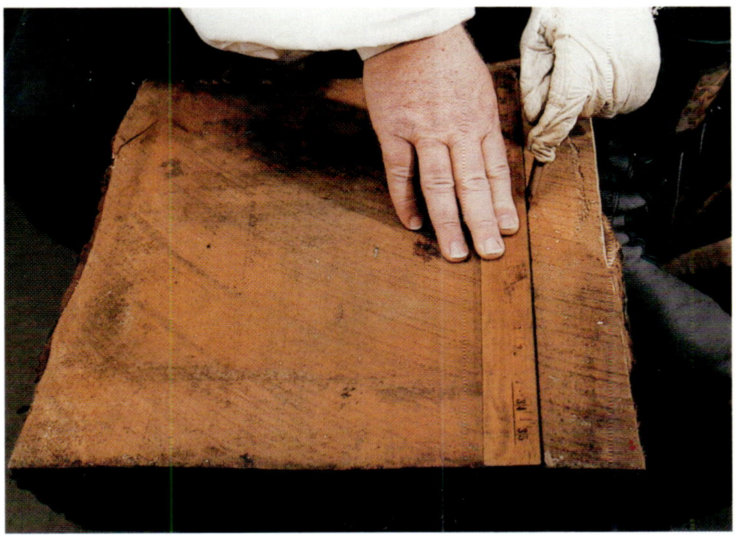

Mark the outside top edge of one 16" side of the bowl. The long side is in the heart wood, just inside the transition line.

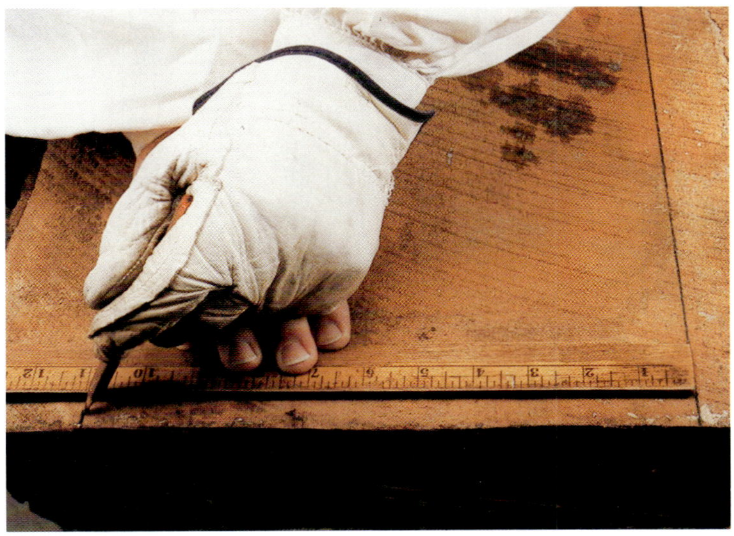

The bowl is eleven inches wide so I measure from each end of the side line I have just drawn...

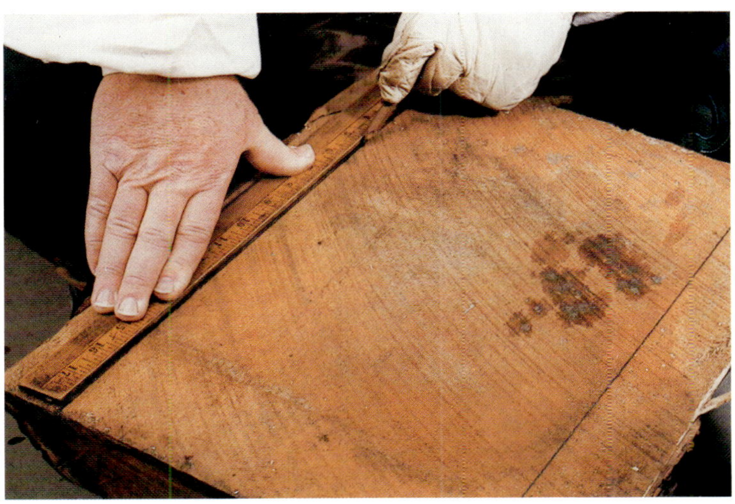

and mark the other long side.

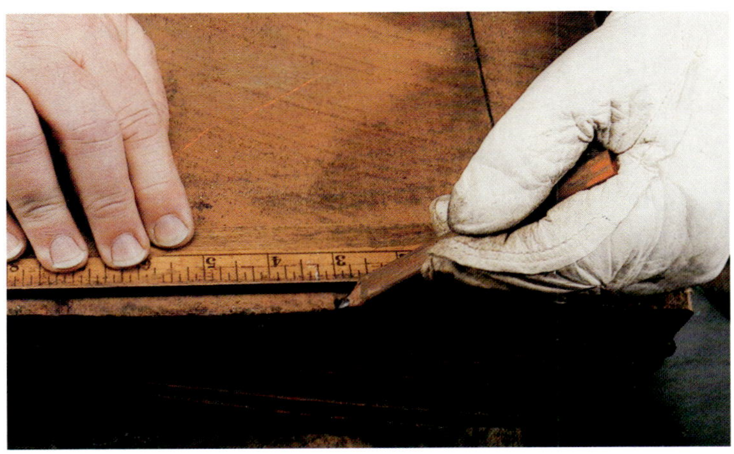

Because the slab is three inches thick, I measure in three inches from the side line and make a mark. Repeat this at each corner.

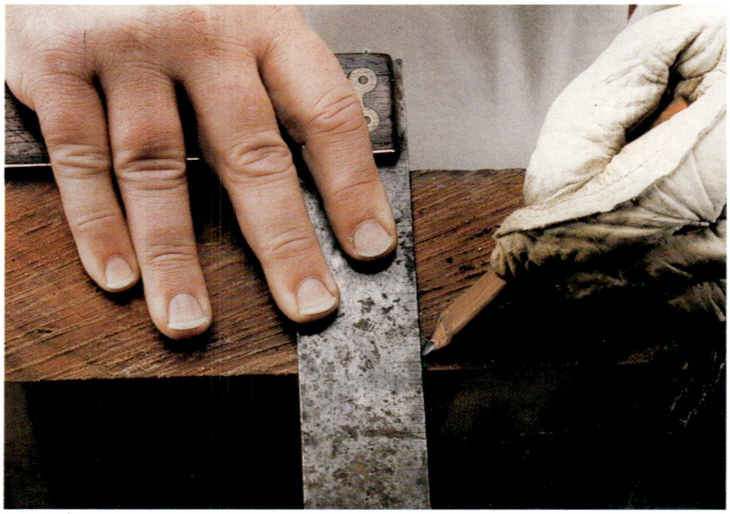

Use a square to carry the mark down to the bottom of the slab at each of the four corners.

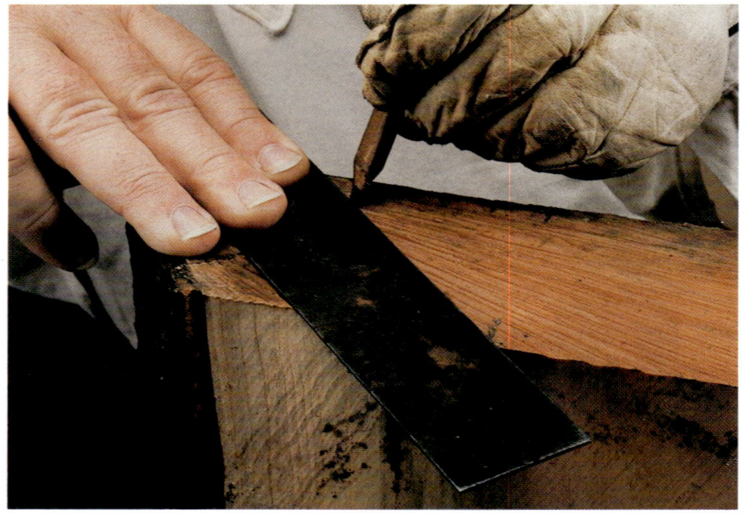

Draw a line from the side line on the top to the mark you have made on the bottom edge of the slab.

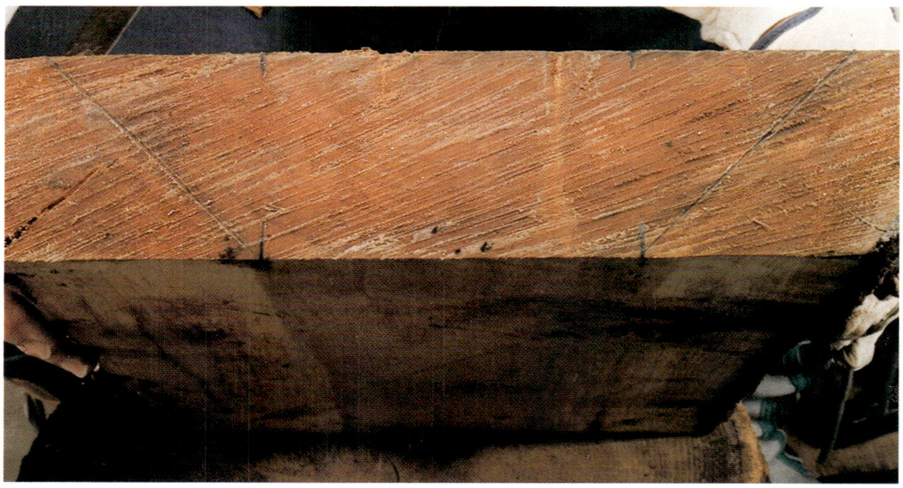

This defines the slope of the bowl.

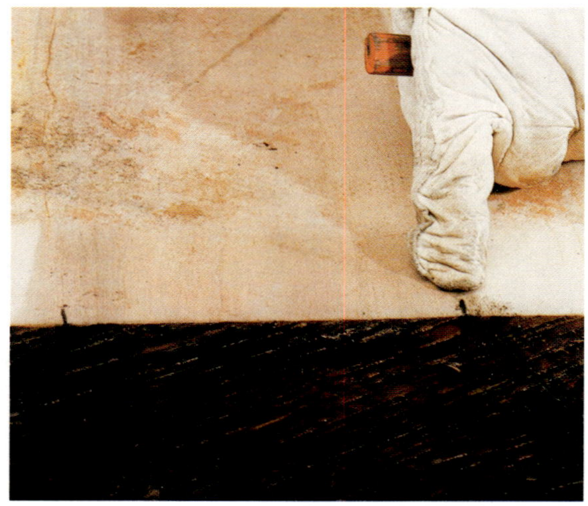

Turn the slab upside down to lay out the lengths of the base. Carry the marks around to the base...

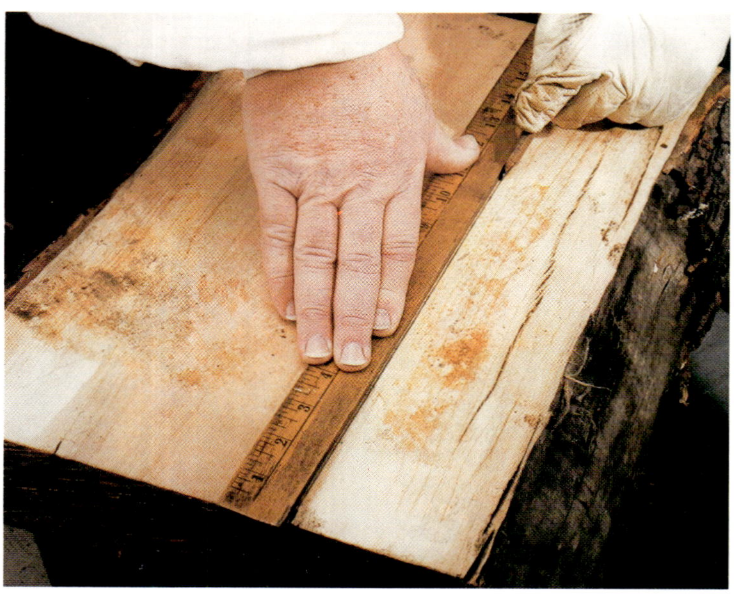

and connect the marks.

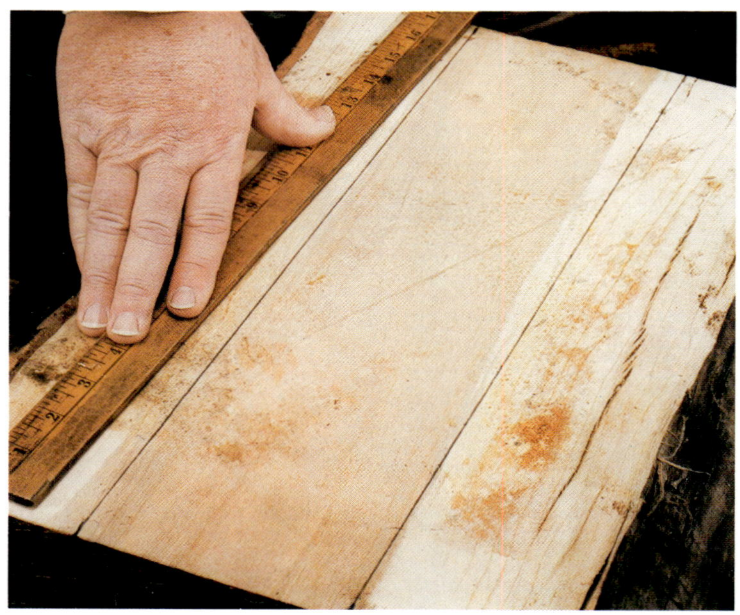

This defines the width of the base.

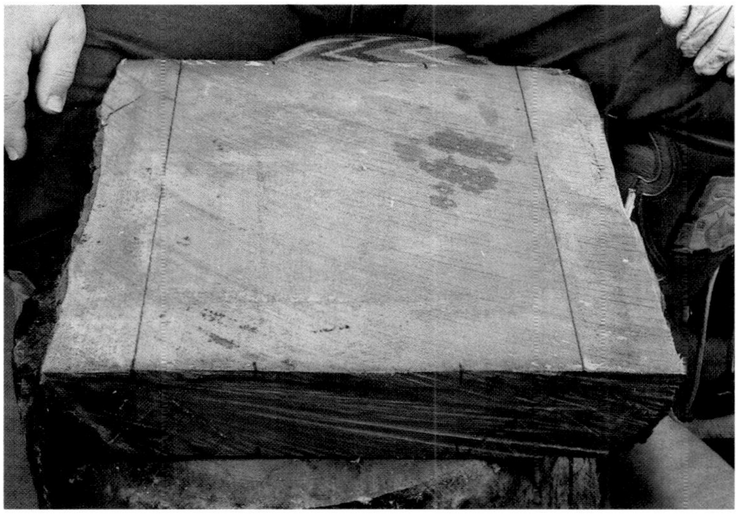

The long dimensions of the bowl are defined.

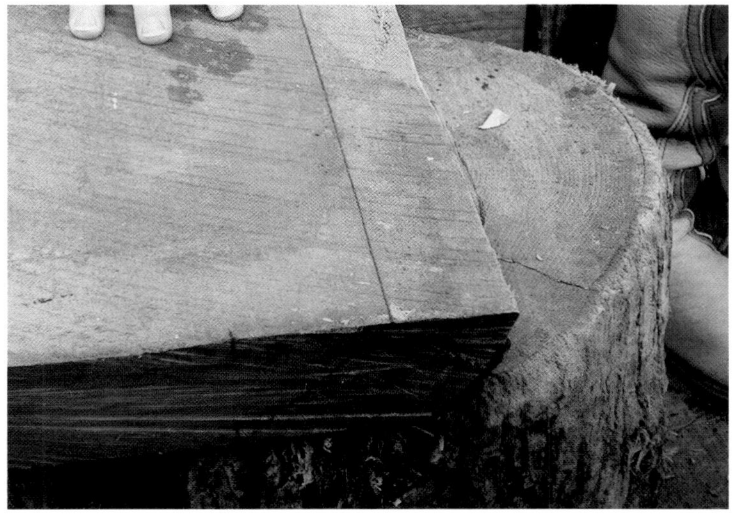

You need a good solid surface to work on. I use a stump about sixteen inches in diameter and 20 inches high. The height will depend on your stool height. Two other safety considerations. First, be sure to protect your eyes from flying chips. Second, I always wear a glove on my working hand. Otherwise I would have blisters on my blisters.

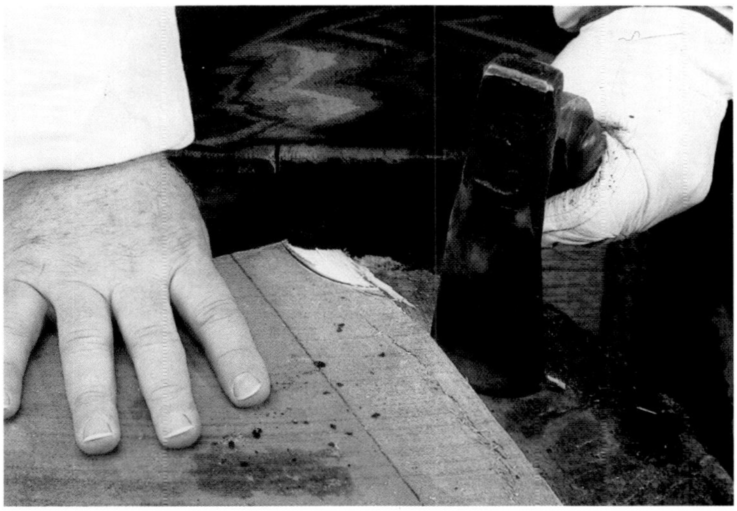

The stump needs to be large enough so that if you miss the adze will go into the stump and not your leg.

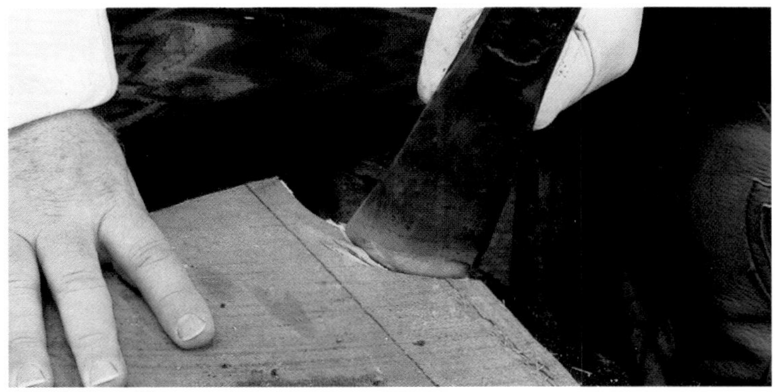

With the slab top side up, work your way...

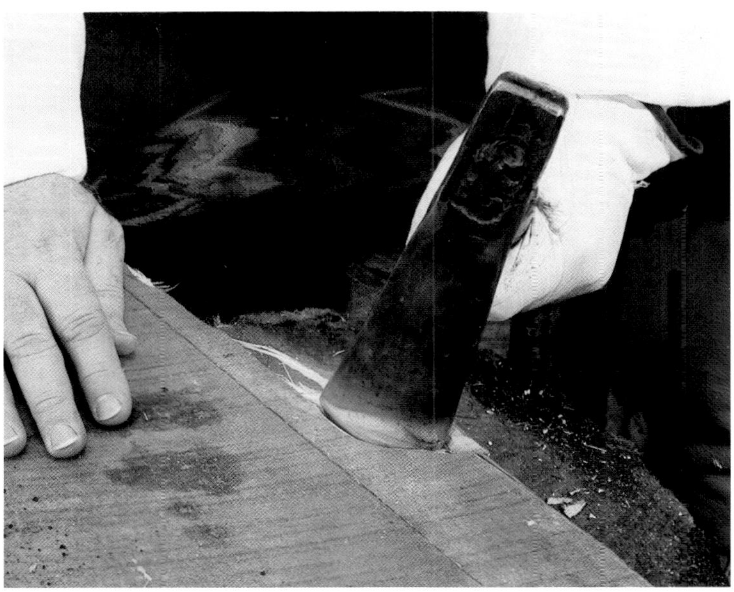

along its edge.

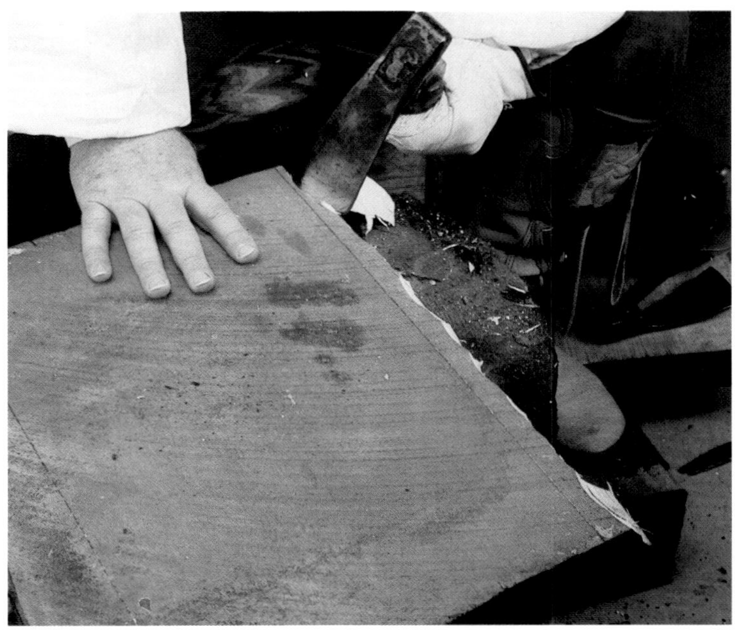

When you've gone the length go back...

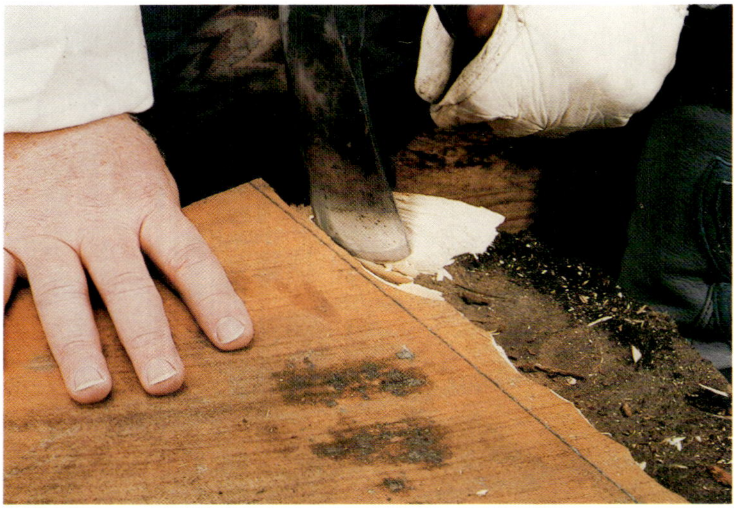

and remove some more, working your way to about 1/4" to 3/8" from your pencil mark.

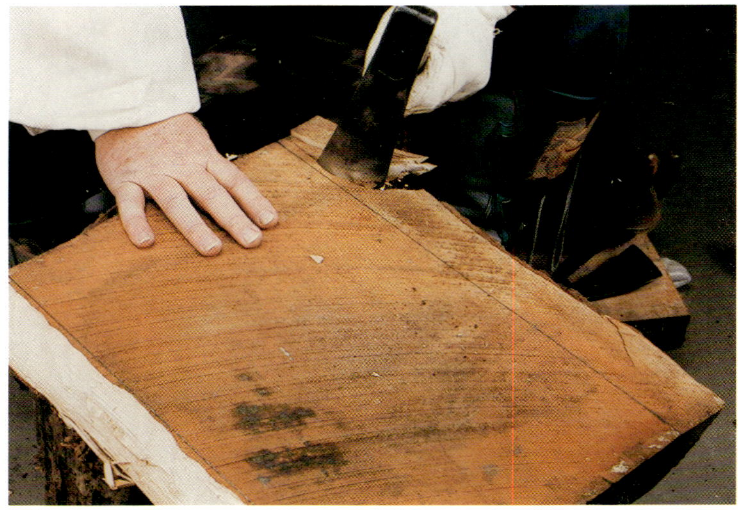

Turn the slab around and follow the same steps on the other side. It is important to keep your work balanced. Don't go too far on one side before doing the other. The fibers in the wood can be under pressure, and if you don't keep the pressures balanced you are more likely to have splitting problems while you're working on the bowl.

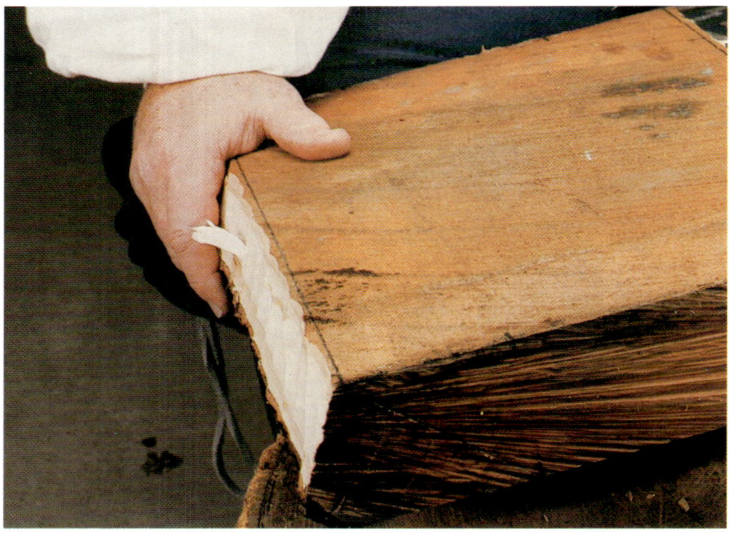

Progress. Leave about 1/4" to 3/8" of the wood outside of the line.

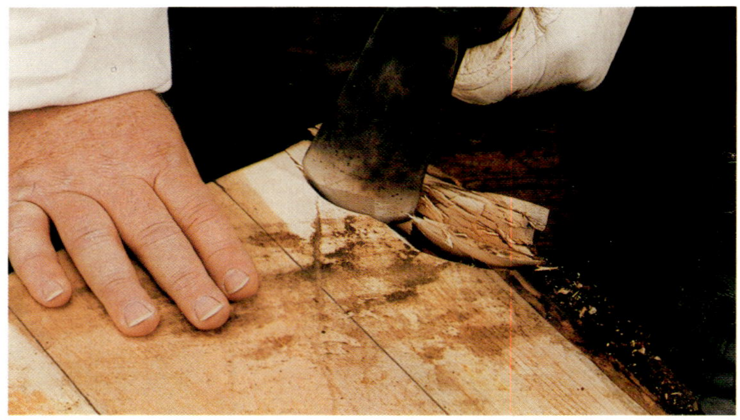

With the top edges knocked off, turn the slab over and begin to shape the slope.

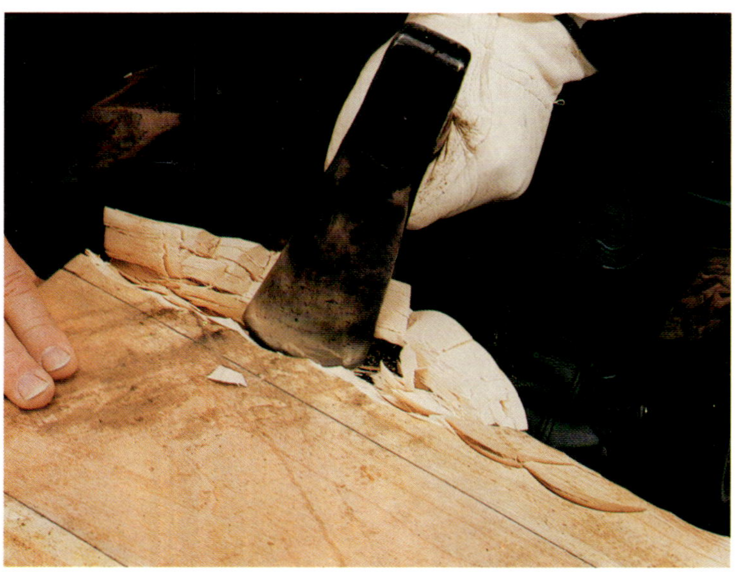

Again start at the edge...

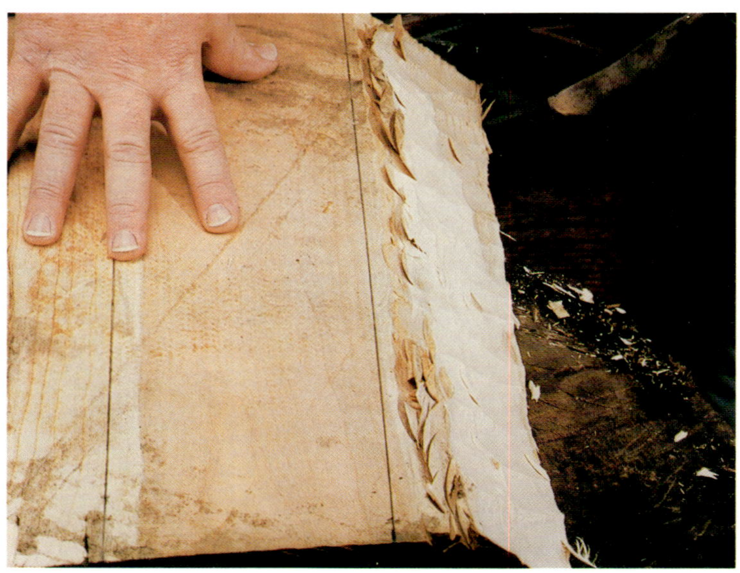

and work your way back to the line.

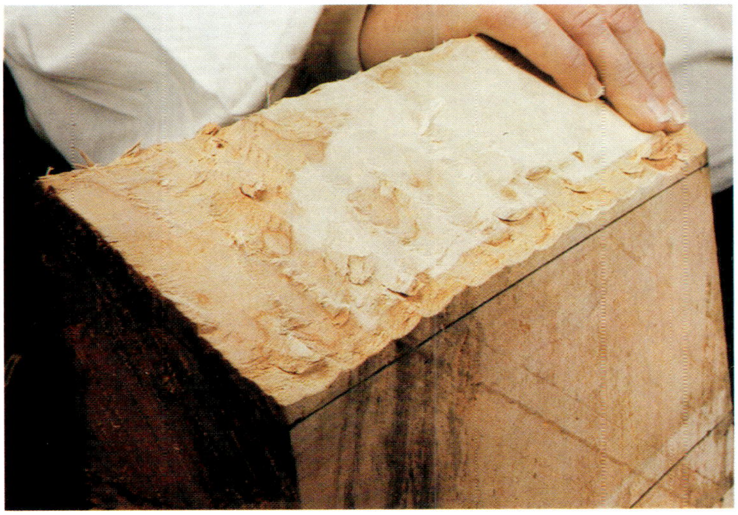

The slope is gradually established.

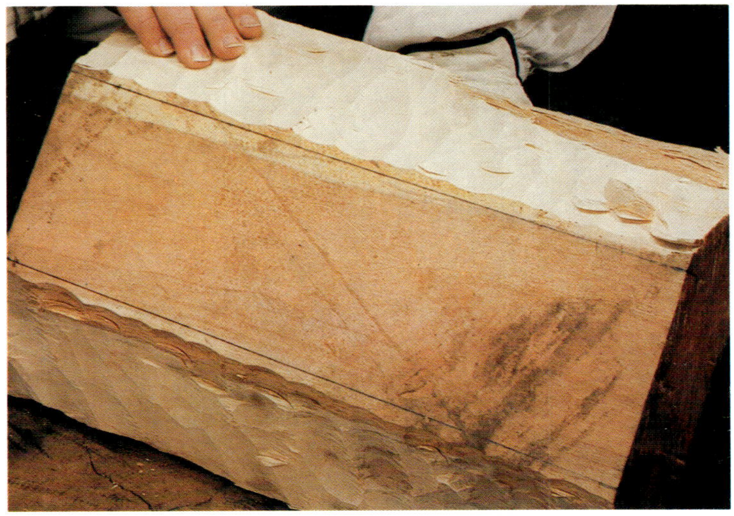

Stop the slope about a quarter inch from the line on the bottom, turn the slab around and repeat on the other side.

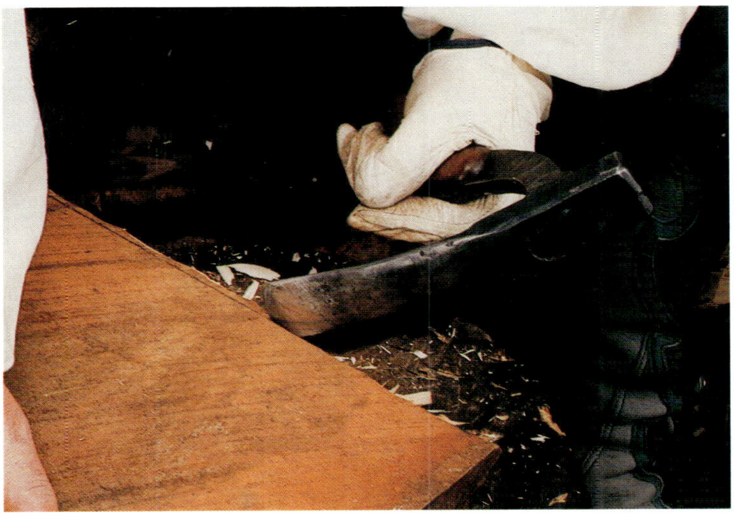

Turn the slab over and run the adze along the top edge to trim it to the line. You need to be aware of the grain of the wood and go with it. If you go against the grain, the adze will go into the wood and you will be tearing it rather than cutting it.

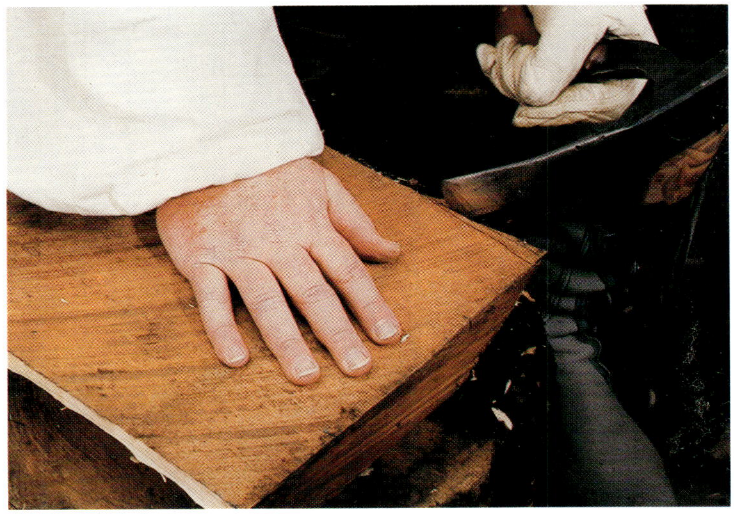

Repeat on the other side.

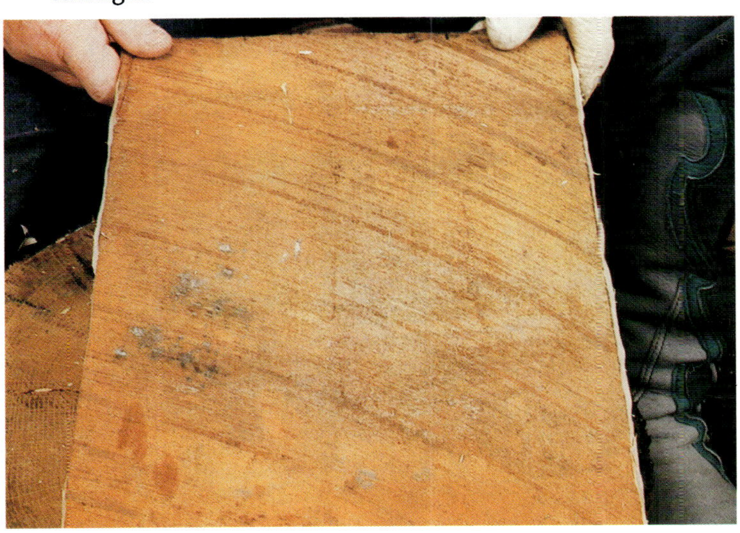

Both edges trimmed to the line. The edge is not yet sharp.

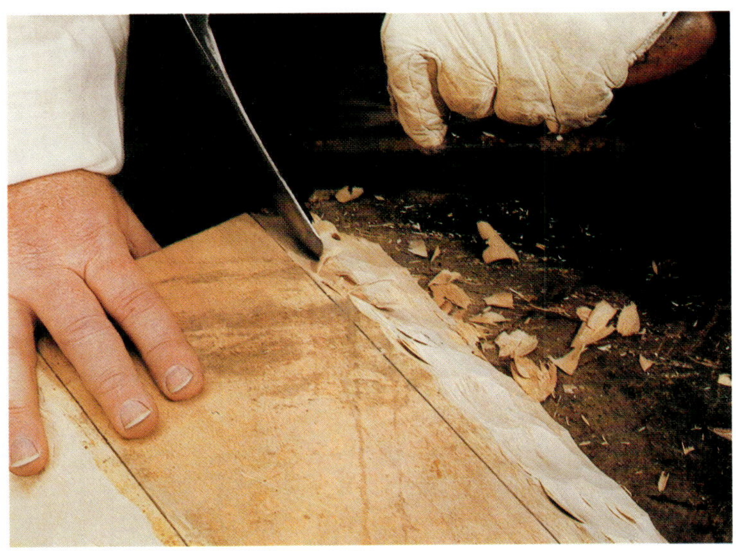

Turn the piece over and bring the slope back to the line on the base of the bowl. Again you will need to go with the grain. In this case it means that I have to "backhand" my strokes, working the adze away from me.

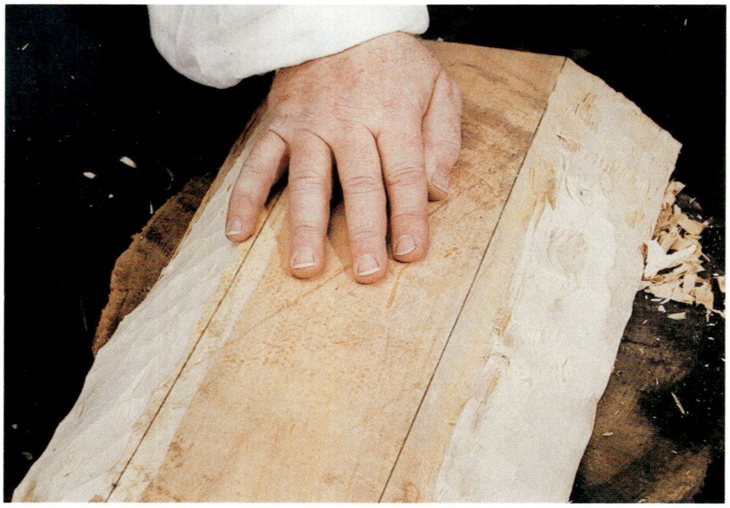

With the line established I can continue to shape the slope.

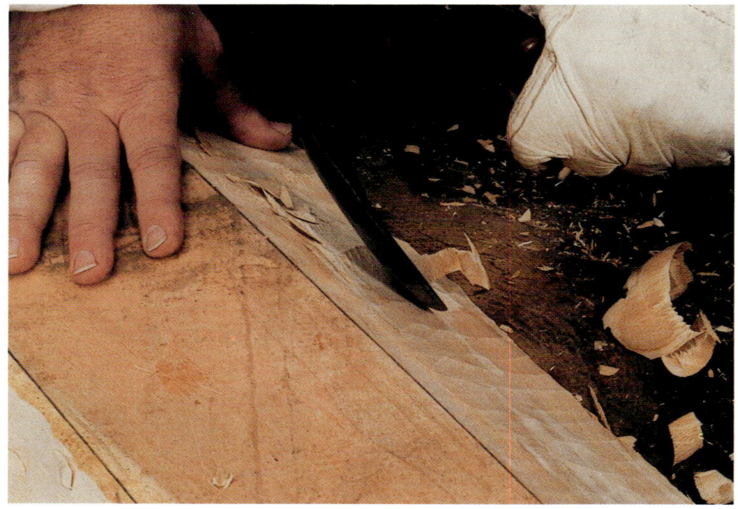

The result is a smoother surface.

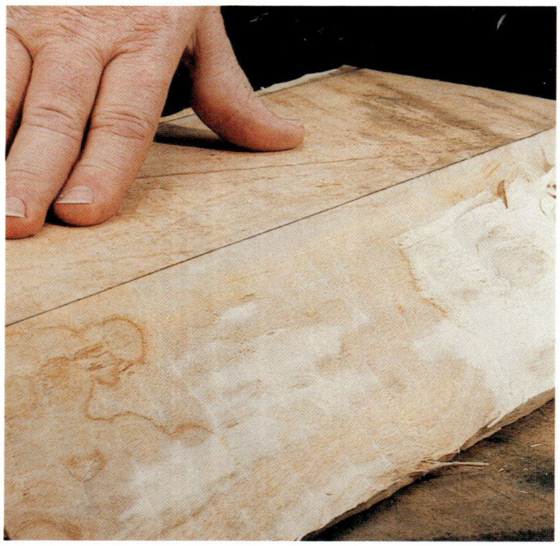

The final roughing of the slope will take it to about 3/16" of the base line. This step should take small chips.

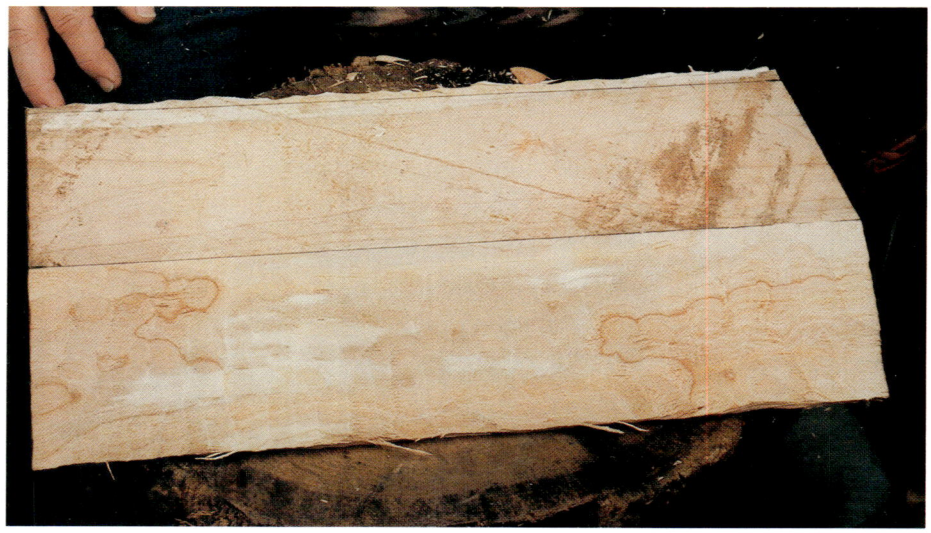

One side complete. This is the last step before the finish cut. Turn the slab around and repeat on the other side.

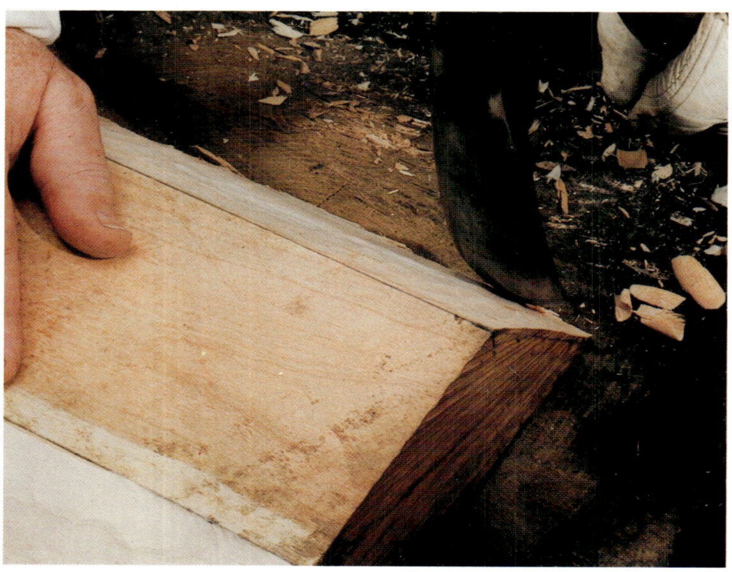

The final step takes the slope to the line and will feather the top edge so that it is sharper.

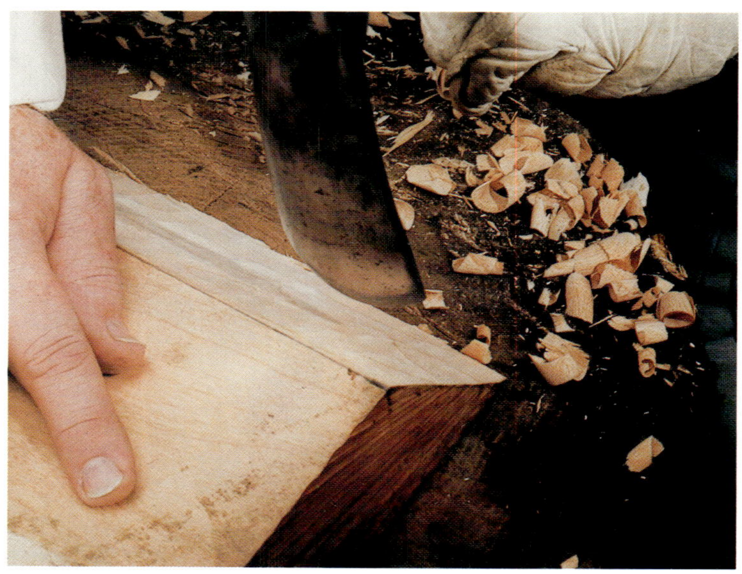

A shorter stroke and softer touch producing small chips, leads to a smooth surface.

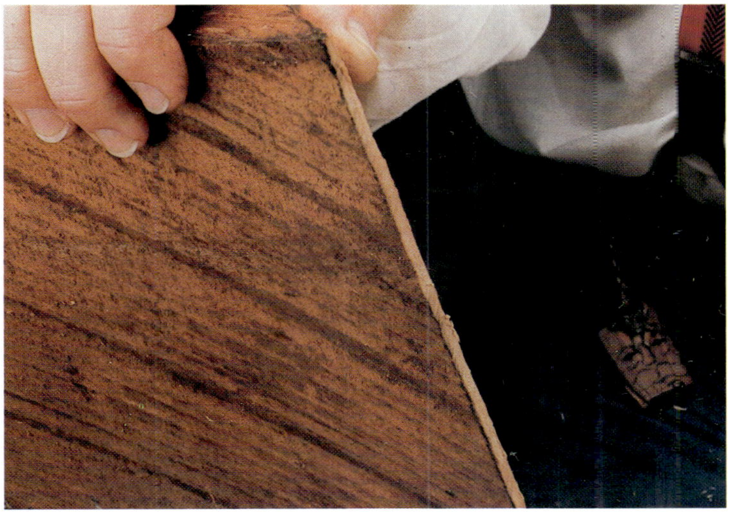

The top edge should be a little less than 1/4" thick when you are done.

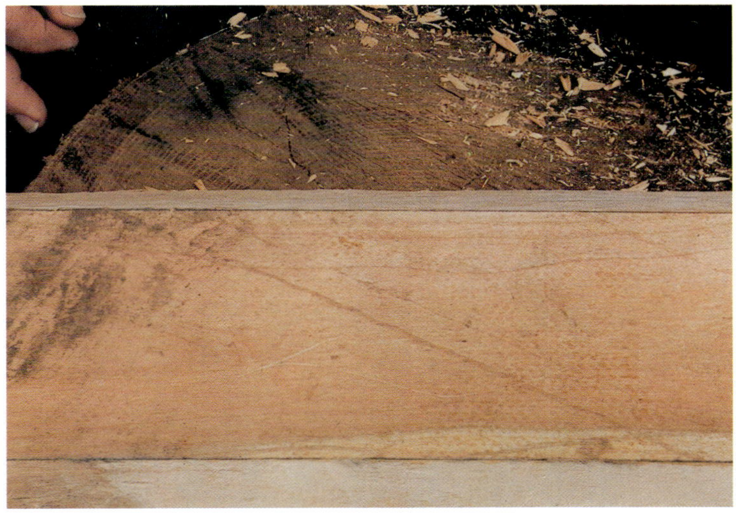

You can check the smoothness and straightness of the slope plane by running your hand over it and by eye-balling it along its surface.

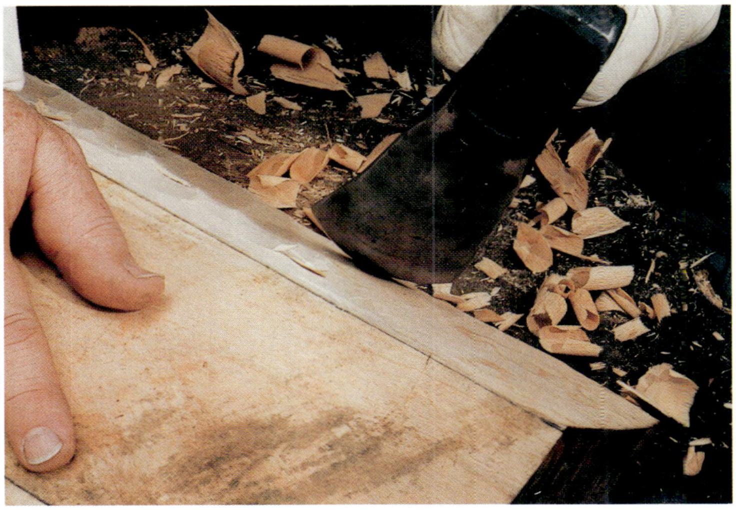

Turn the slab around and repeat on the other side.

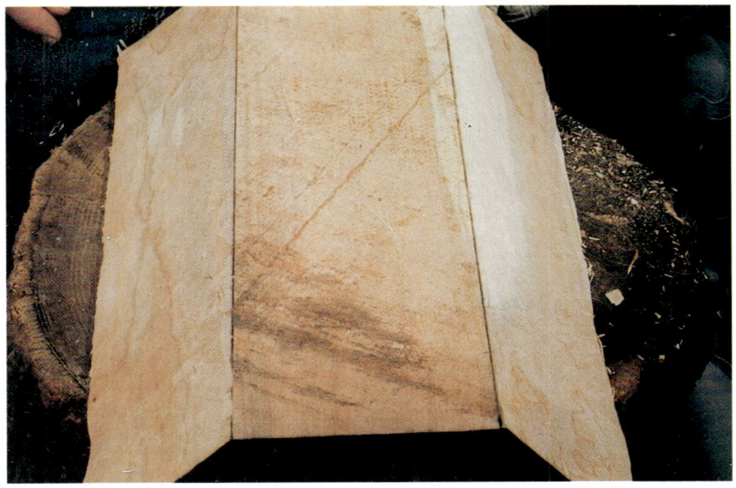

The slopes are established.

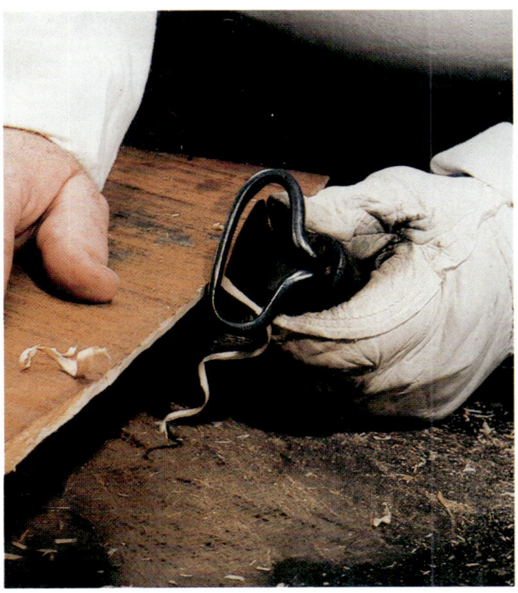

I'm using a bowl shave to smooth the top edge. Be sure to go with the grain. A rasp or plane will do the same job.

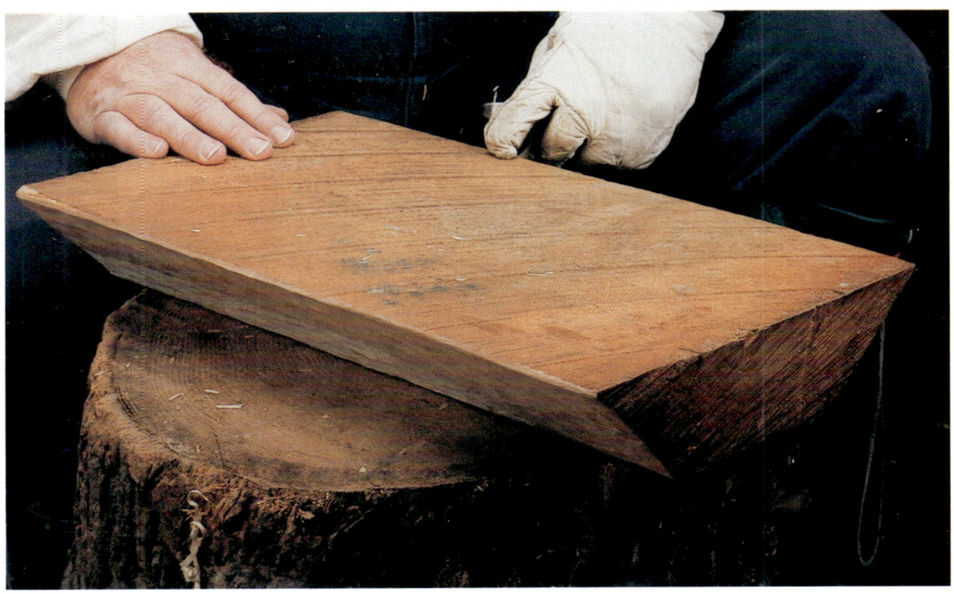

The edges are straight, the thickness of the rim is established, and we are ready to work on the ends.

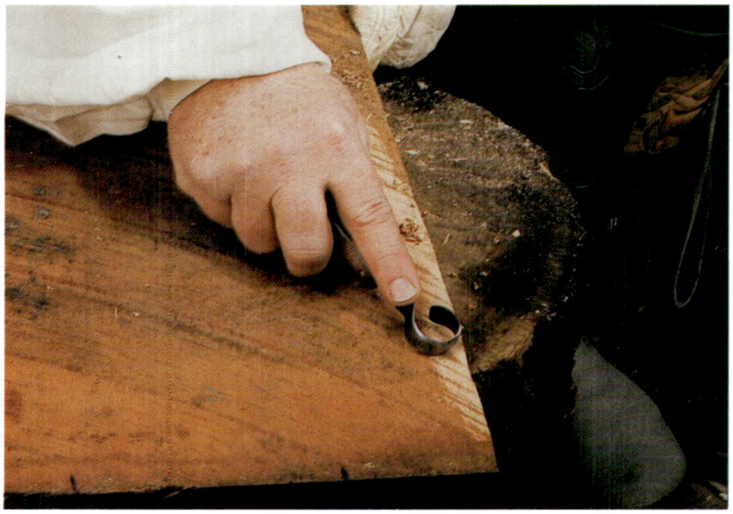

Before marking the ends, I smooth the top surface along the edge using a scorp. This helps me see if there are any splits or knots that will cause problems later. Be sure to go with the grain.

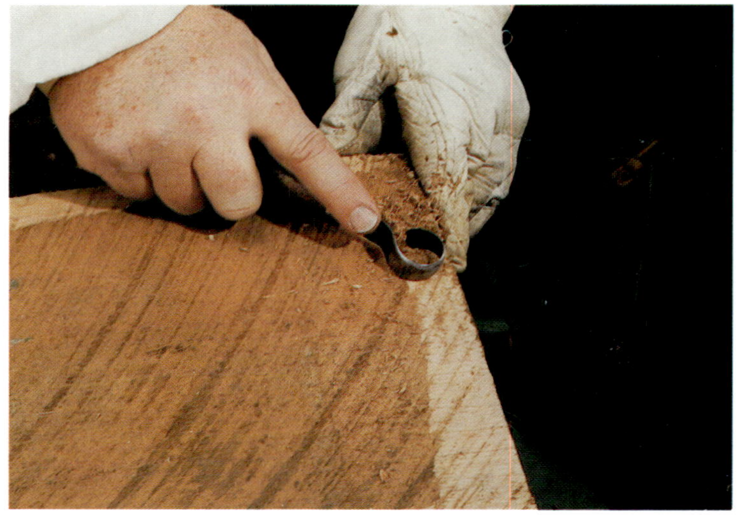

Do the same thing across the end. Because you are going across the grain the scraping will be completely different. It helps to go at a little angle to the grain.

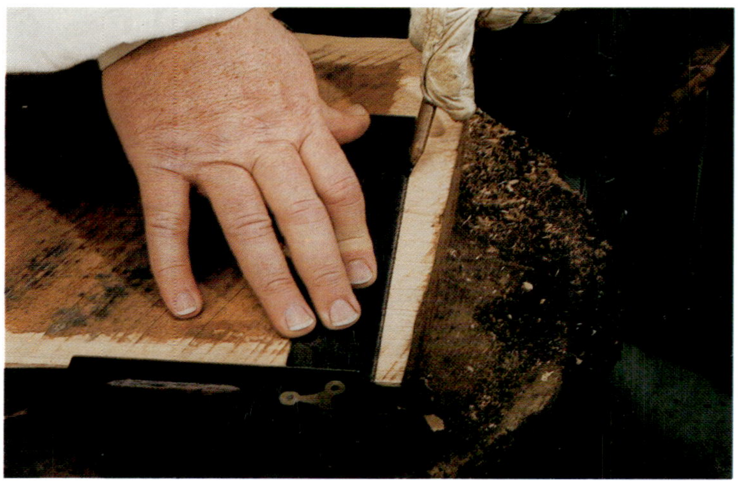

Use a square against one of the finished side edges and draw the end line. Carry it across.

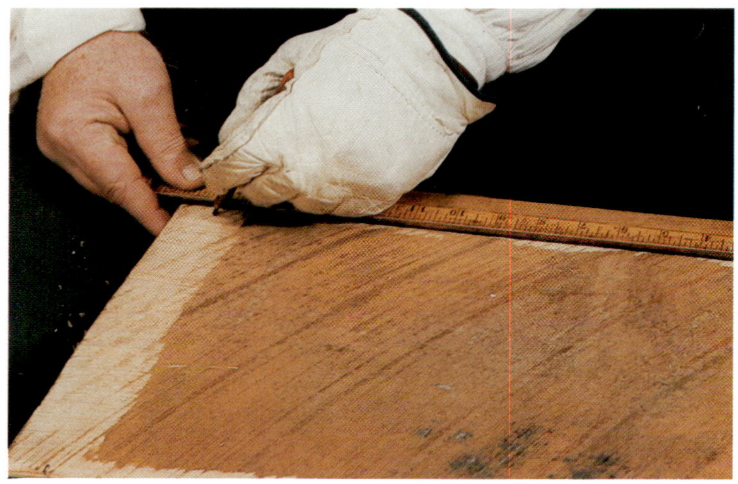

The bowl is to be 16 inches long so I measure from the end line 16 inches along each edge and mark.

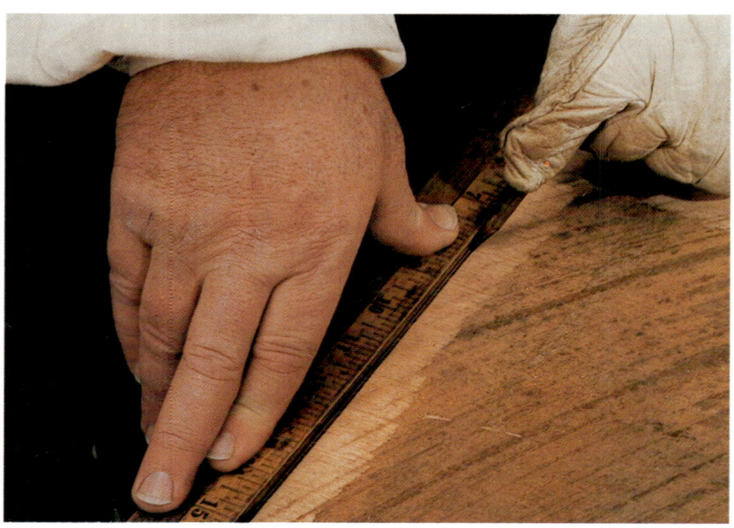

Draw a line between the marks to establish the other end line.

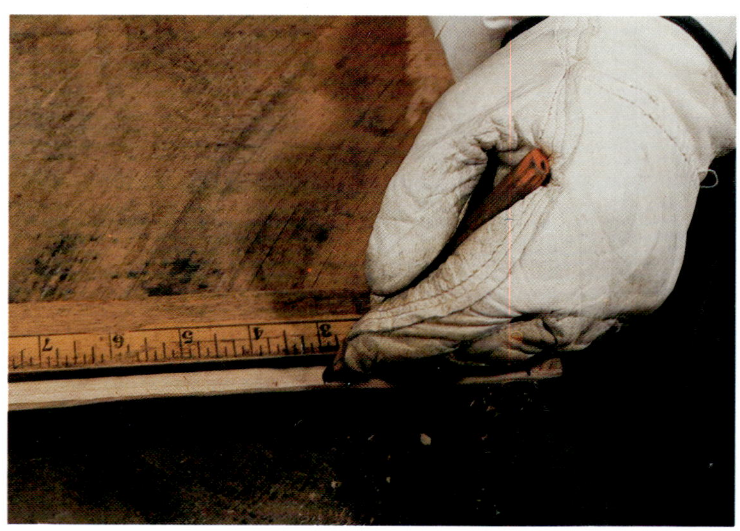

On the edges measure in 3 inches from each end.

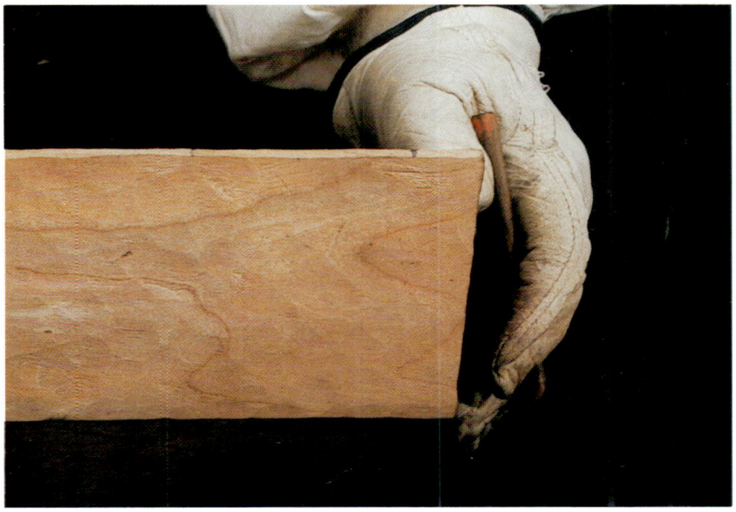

Carry the marks around the side.

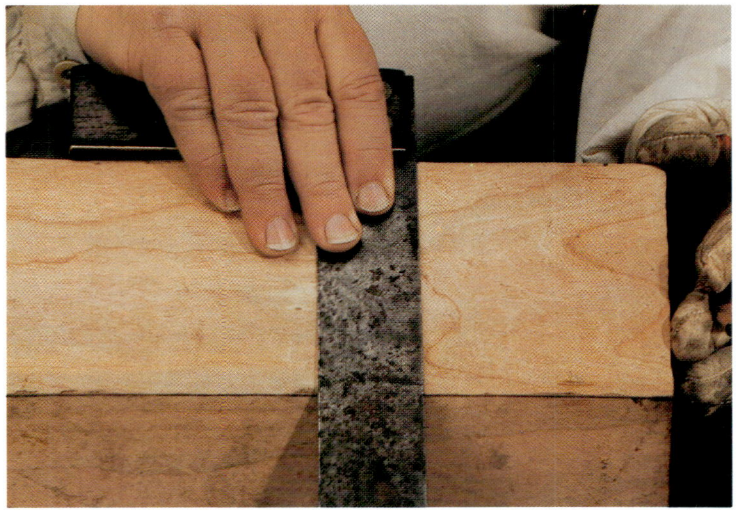

Use a square to carry the 3 inch mark to the bottom.

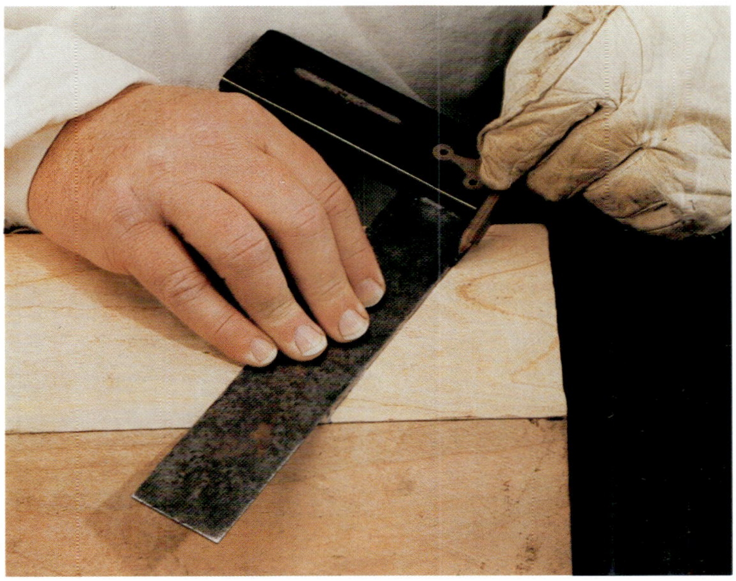

With a straight edge draw the line from the 3 inch mark on the base to the end mark at the top. Repeat at each corner.

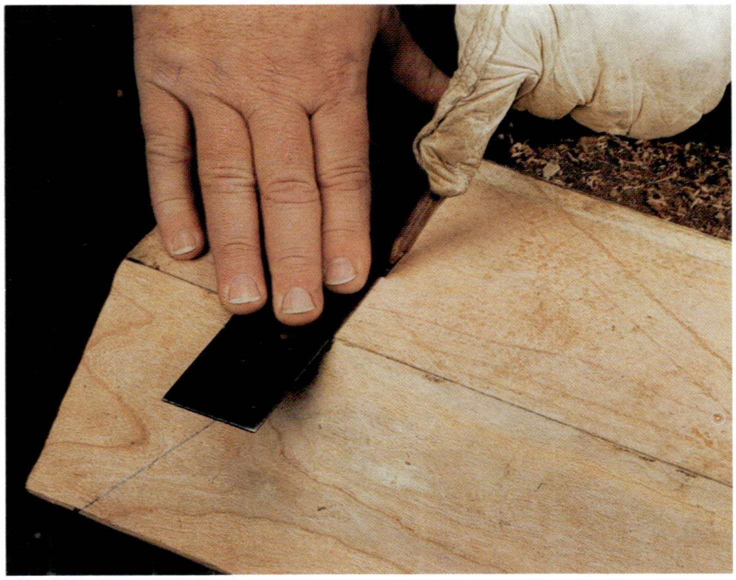

Connect the 3 inch marks across the bottom.

The lines of the base are established.

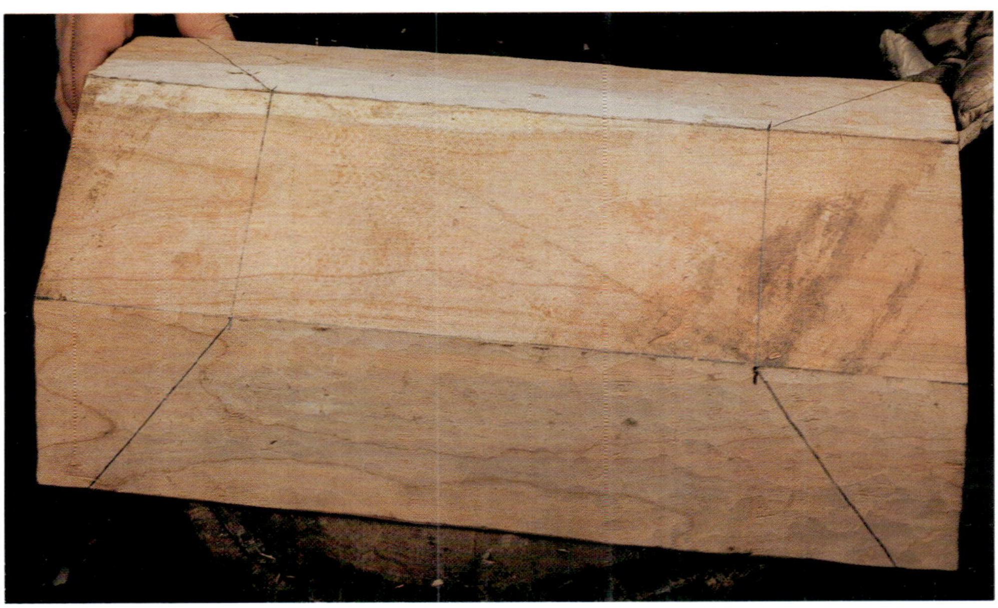

Hewing cross-grain, like here, is much harder. Make sure the surface under the slab is flat and clean. Use small strokes, working from the end toward the line.

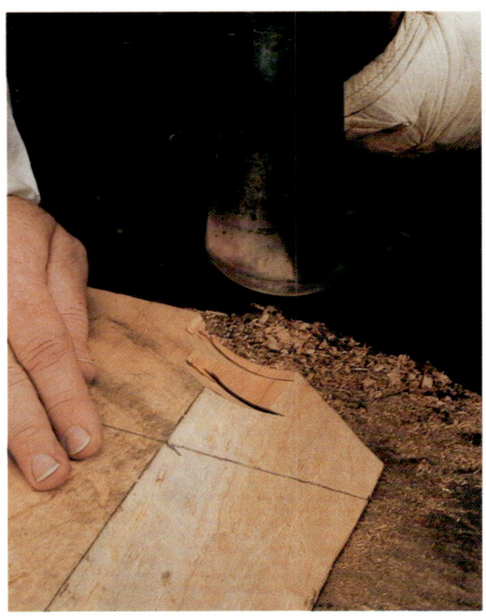

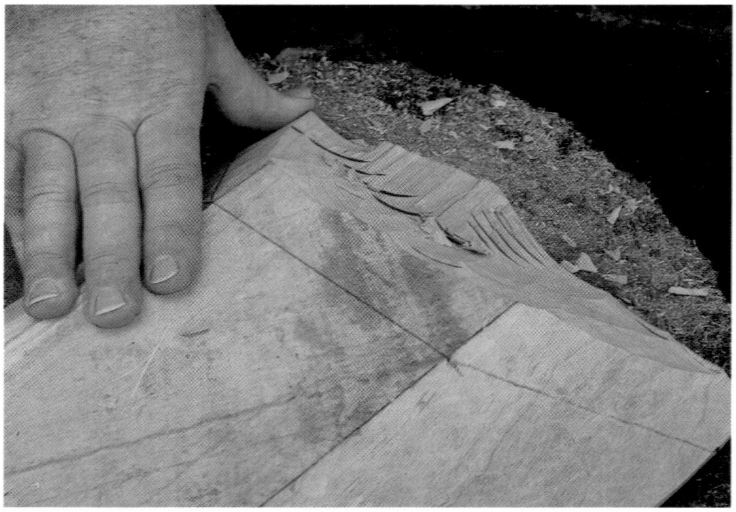

Work from side to side.

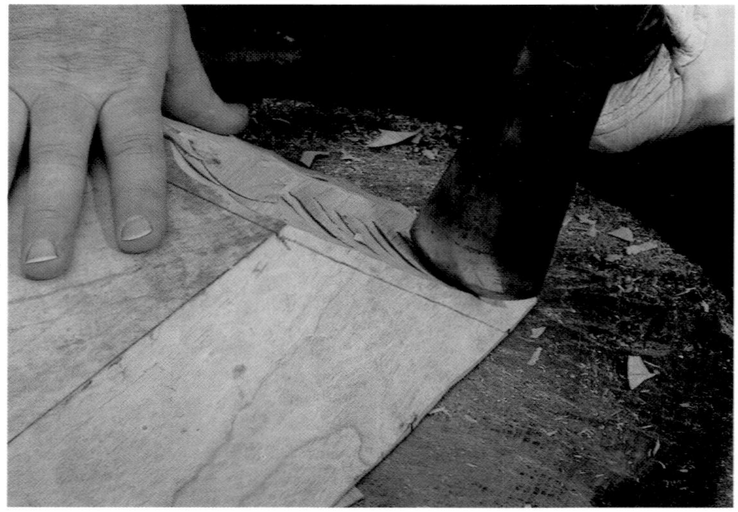

You'll want to come about this close to the line at the this stage.

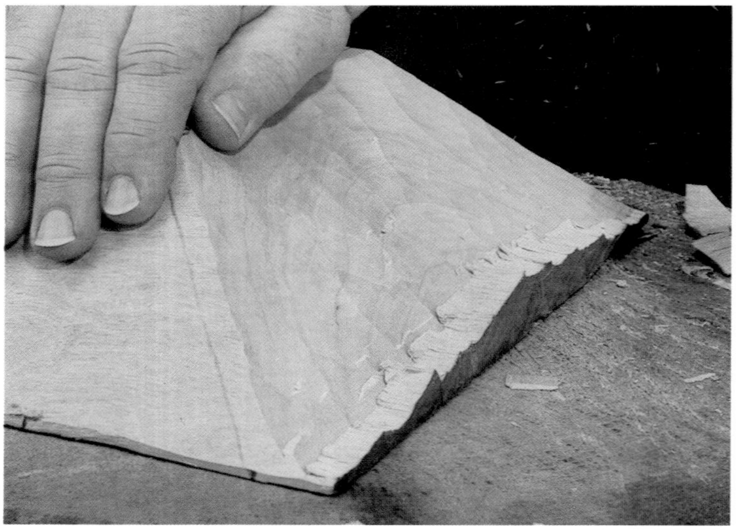

When hewing cross-grain, it very important that the piece be tightly against the working surface to avoid chipping the top rim of the bowl.

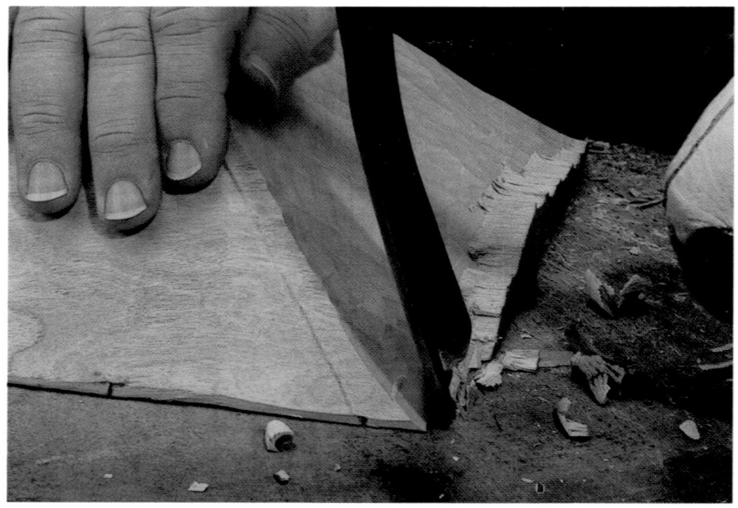

Work your way across the end.

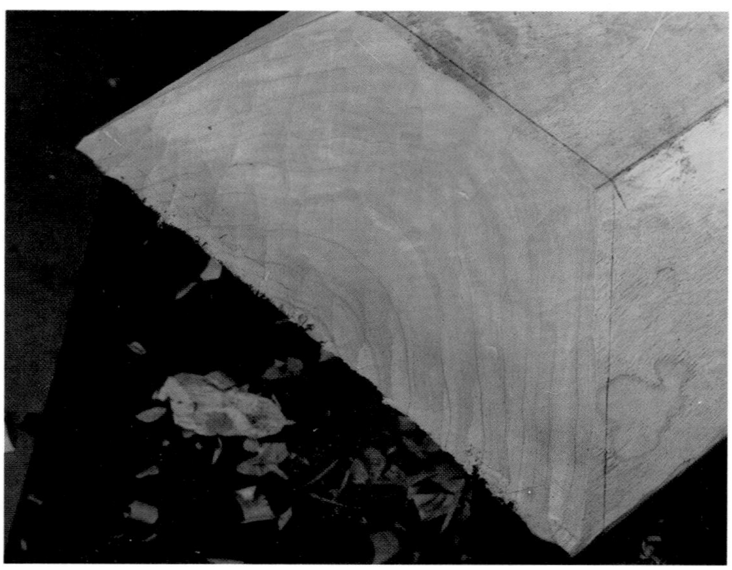

Progress. Turn the slab around and repeat on the other end.

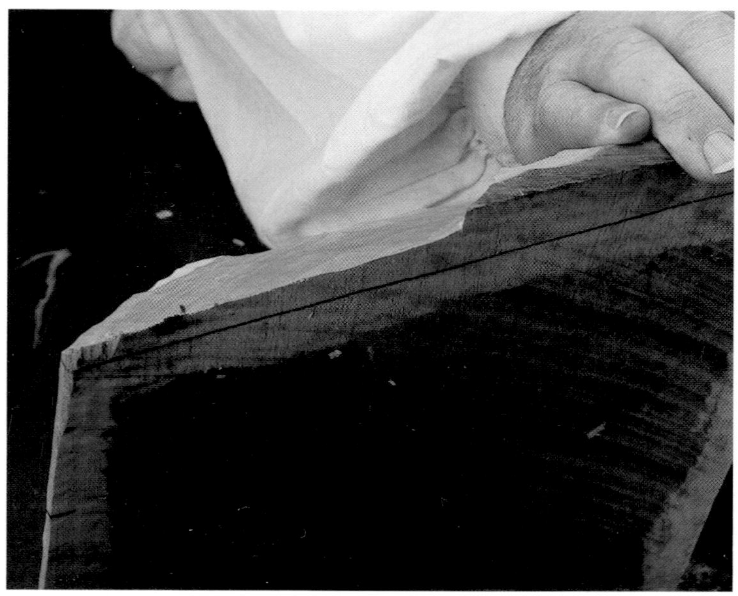

Pick up the slab and check occasionally to be sure you stay outside of the end line on the top of the bowl.

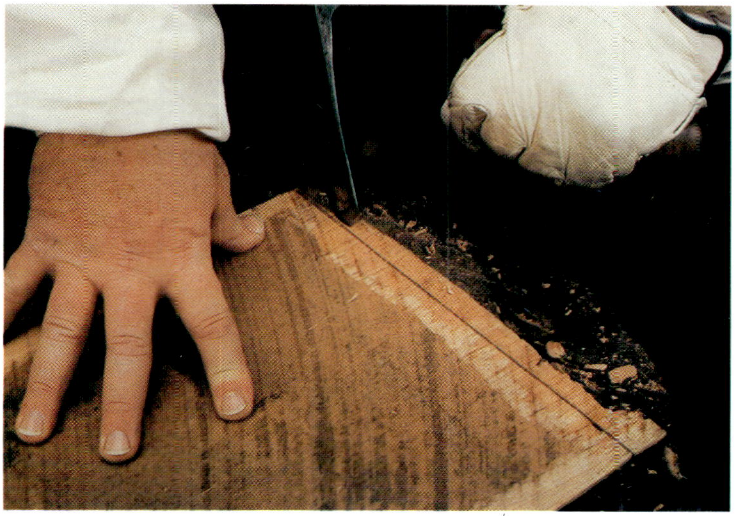

Turn the piece over and begin to work the end back to the line. You need to work in from both corners to avoid splitting. Come from one side...

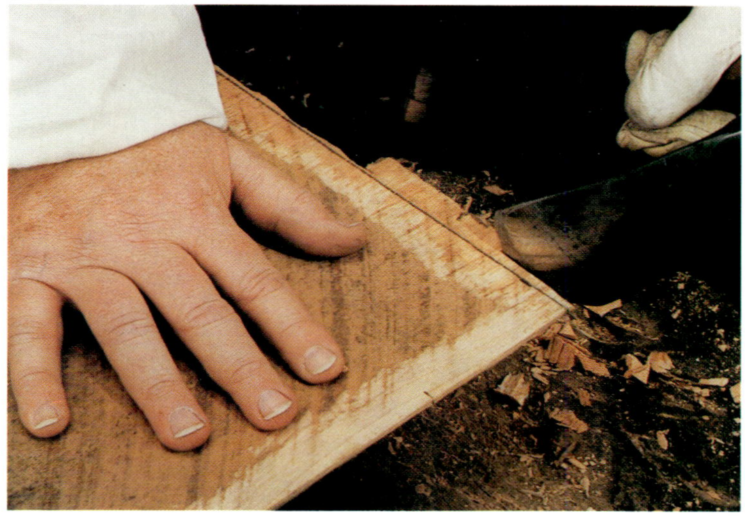

and the other.

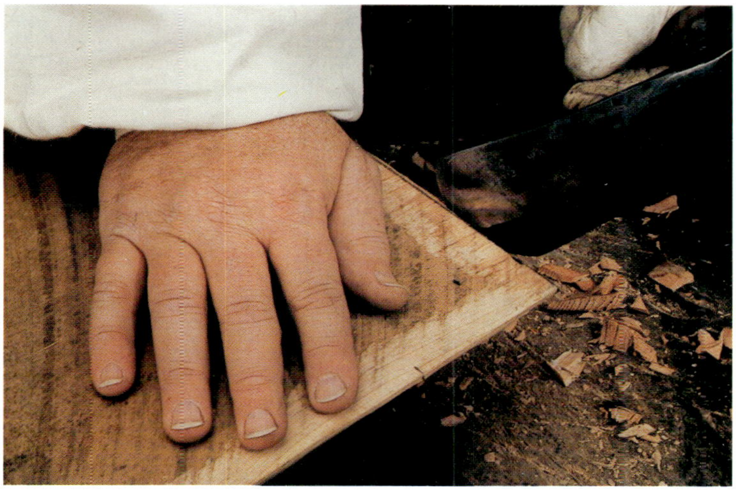

After roughing the edge go back to refine it, bringing it to the line. Again work from the sides toward the center.

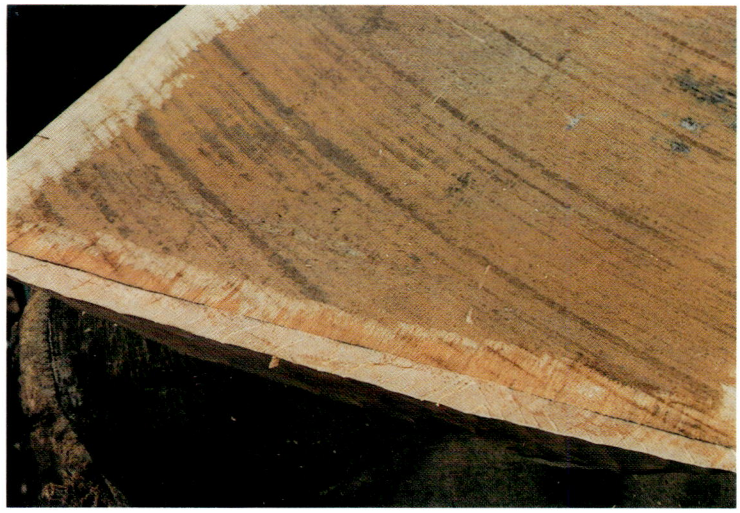

One edge refined. Repeat at the other end.

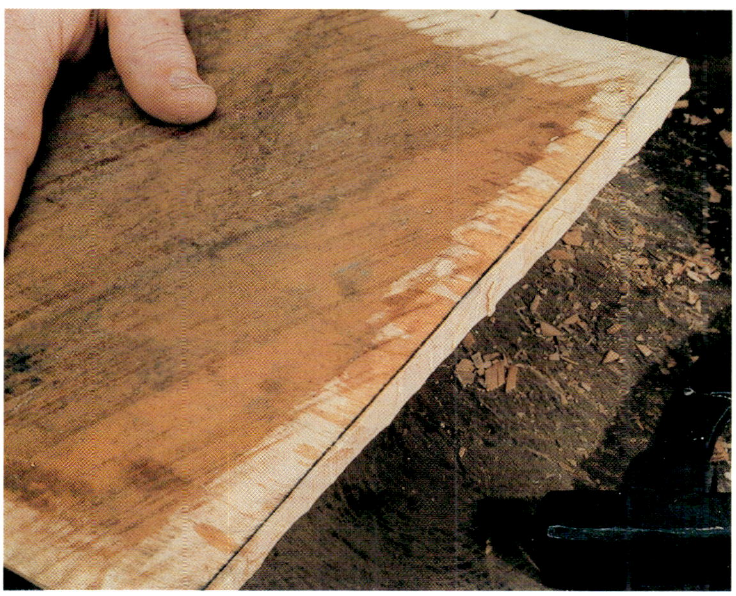

At the other end it is a little thick for refining...

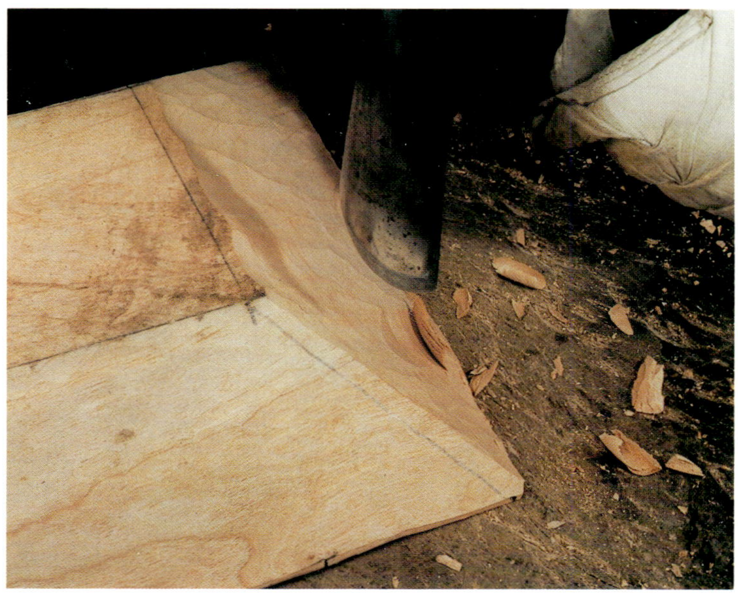

so I turn the piece over and take some off the edge of the slope.

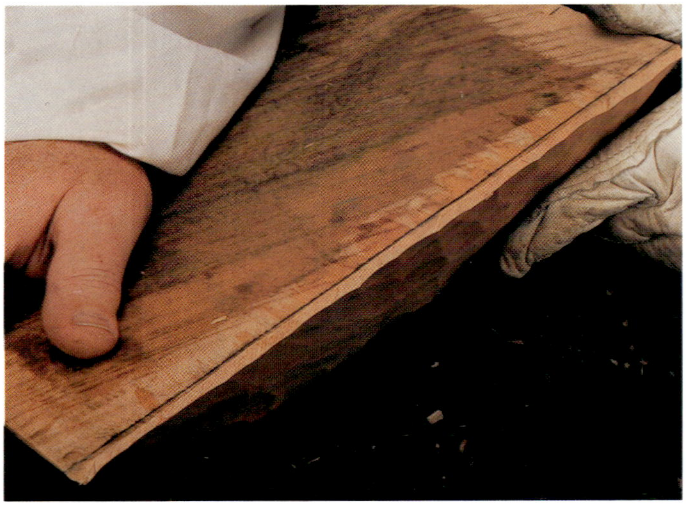

This thickness will be much easier to shave.

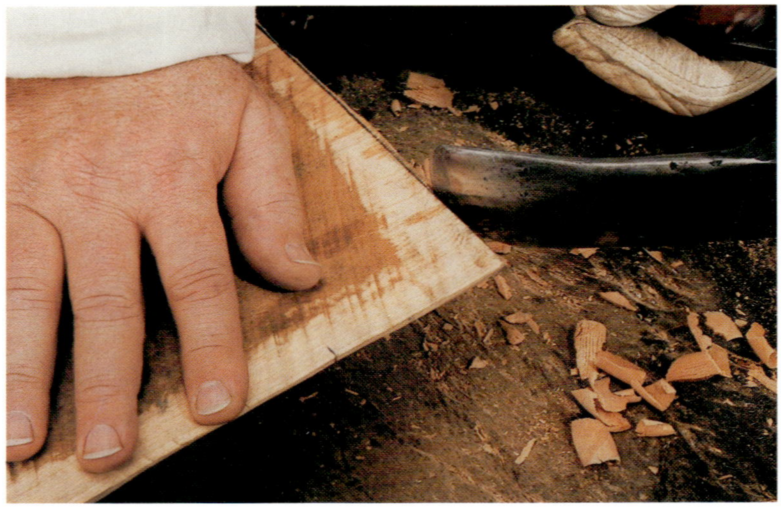

Now continue trimming to the line.

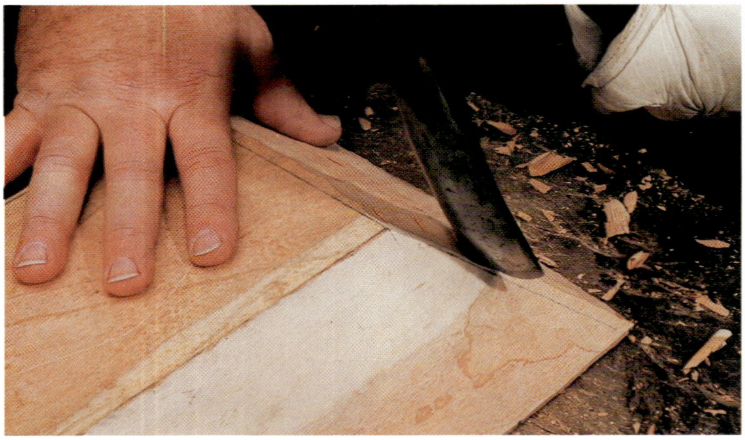

Take the end slopes to the line at each corner.

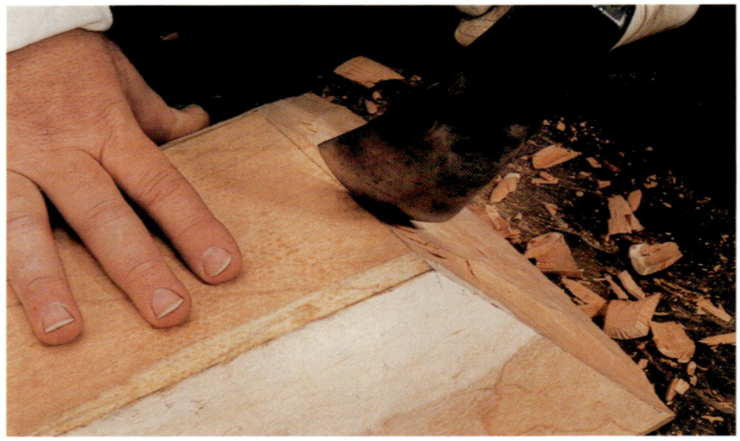

Then go across the base to the line.

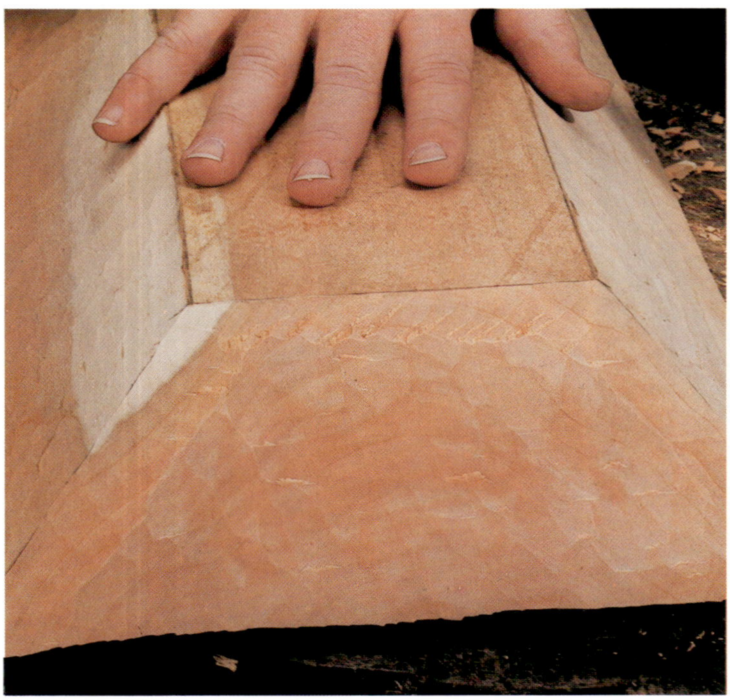

This takes you to this point.

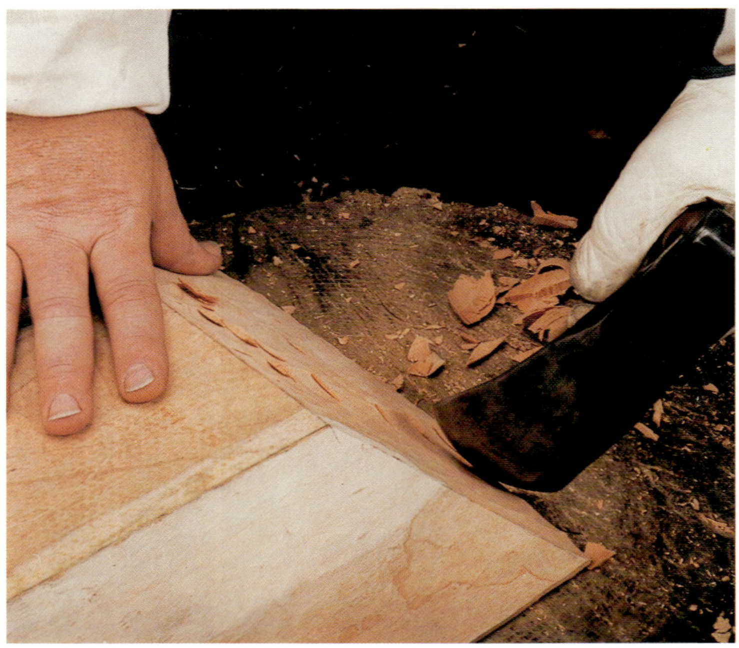

Remove the center of the end slopes cutting at an angle across the grain, and working your way from the edge to the center.

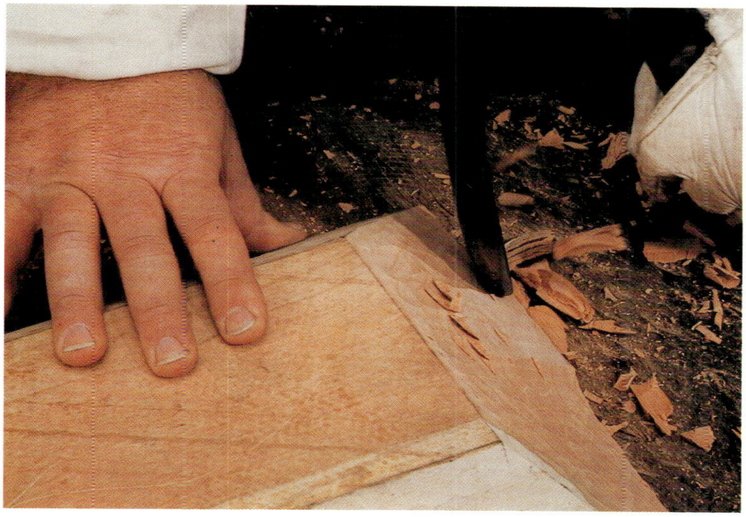

Come back from the other side.

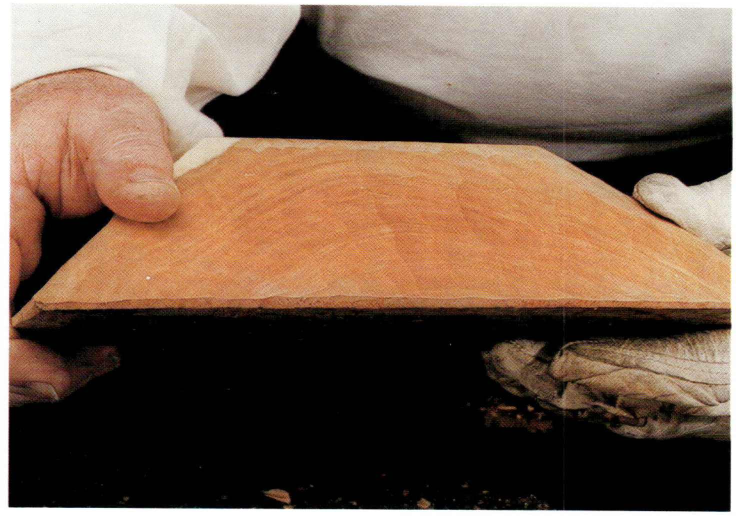

The rough cut takes you down to about the thickness you want for the rim.

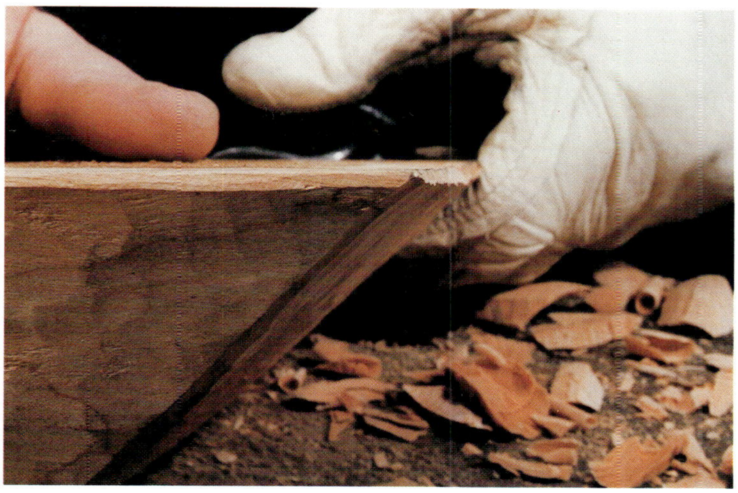

The rim at the ends has a small bevel that needs to be removed.

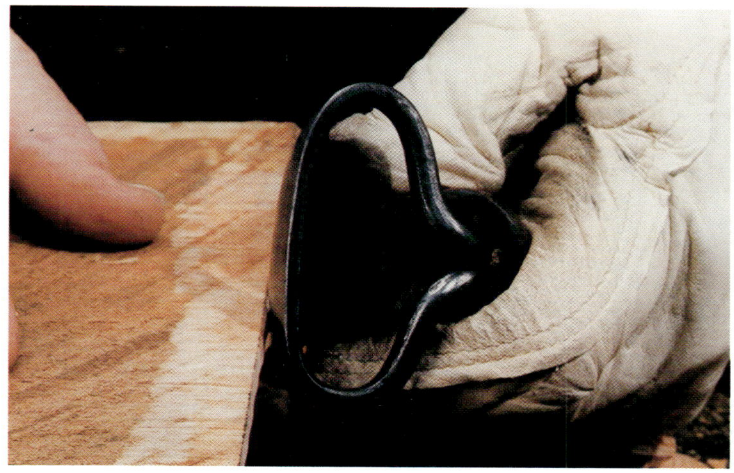

Working in from the corners, plane the ends to make them smooth and straight.

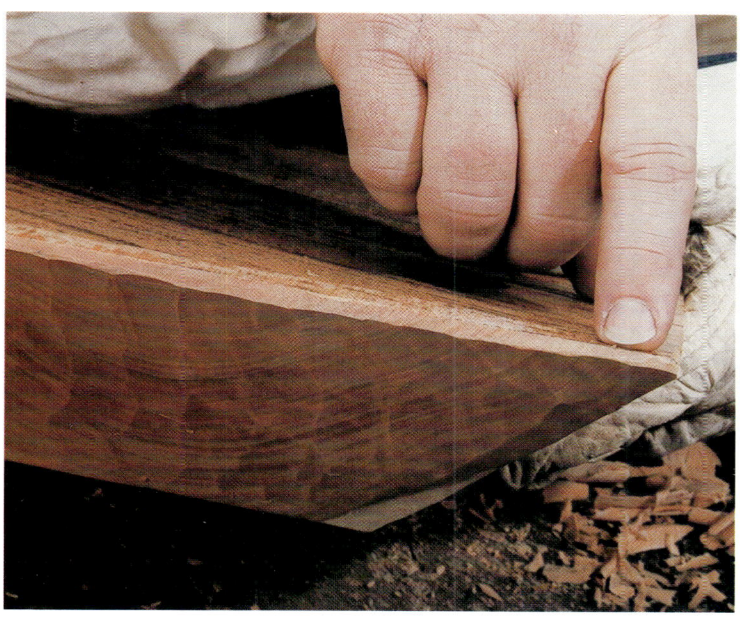

That takes us to this point. I want the whole edge to be about this thickness when its finished.

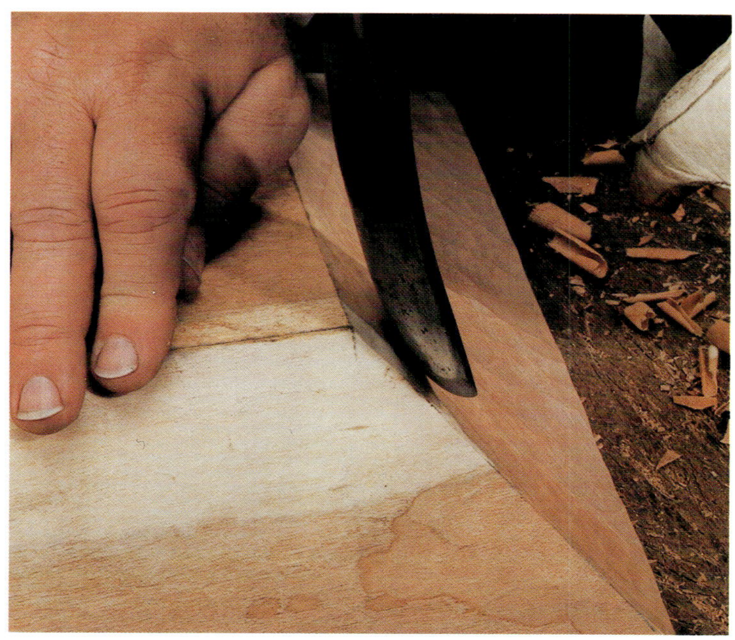

Do a final smoothing and shaping of the ends.

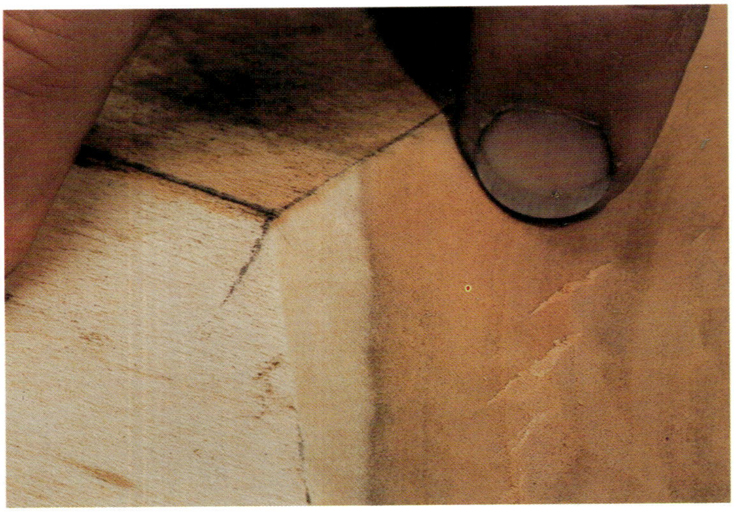

While you are learning to use the adze you may leave chop marks like this.

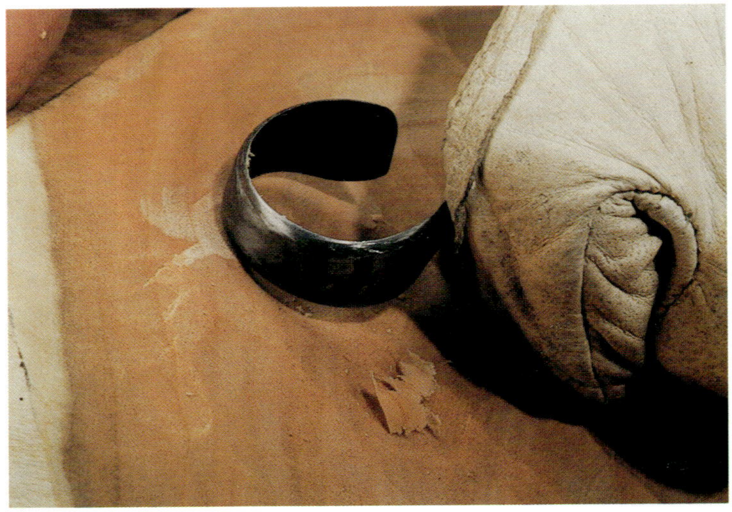

These can easily be removed with a scorp.

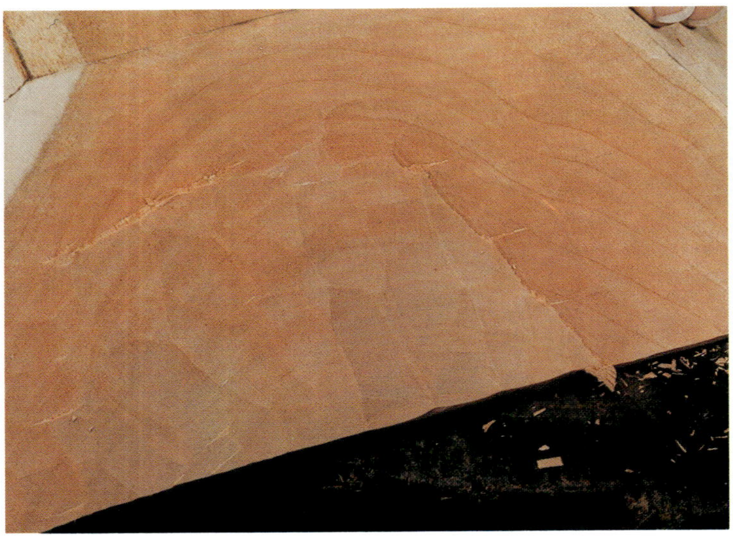

Progress. Remember the finish stroke with the adze takes fine, little chips.

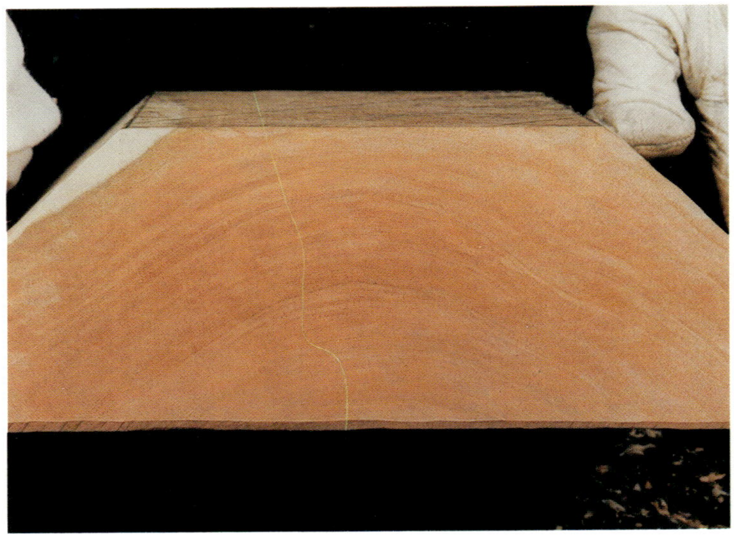

One end complete. Repeat at the other.

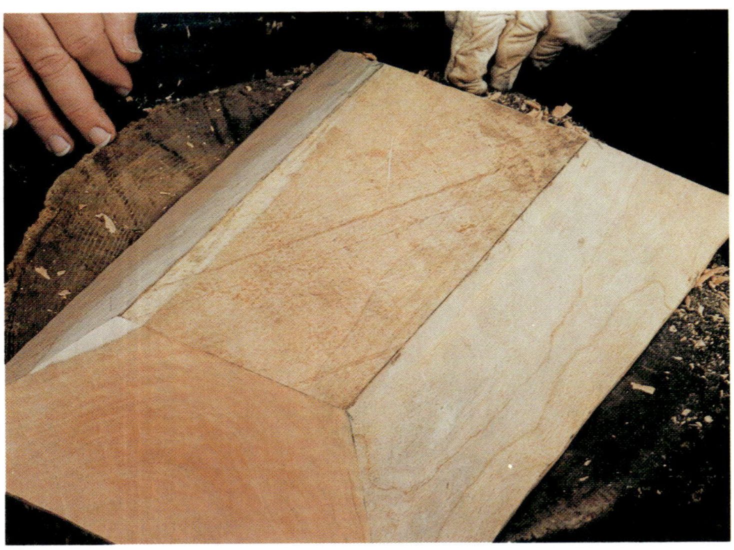

The outside surfaces are now defined. We are ready to mark the corners of the bowl.

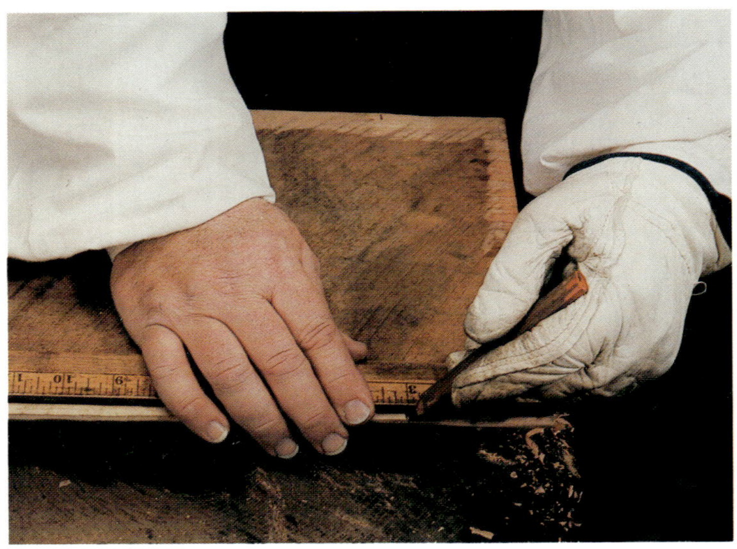

Turn the slab over with the top up. From the corner measure 3" in each direction and make a mark on the edge.

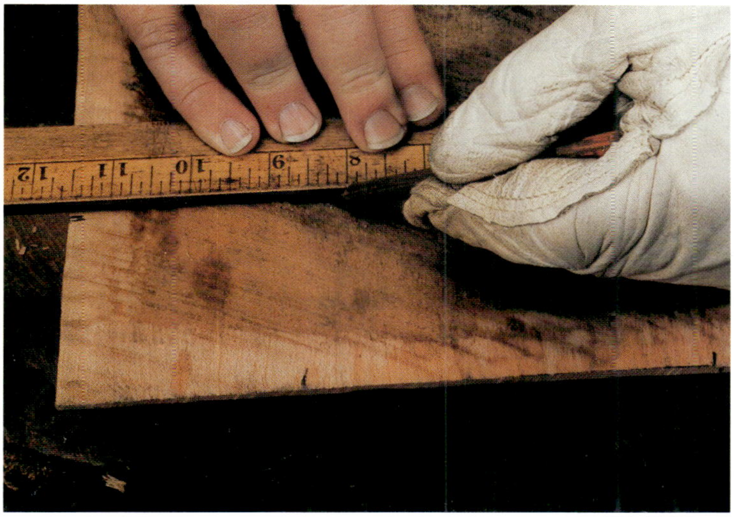

Establish a point for the compass by making a mark perpendicularly in

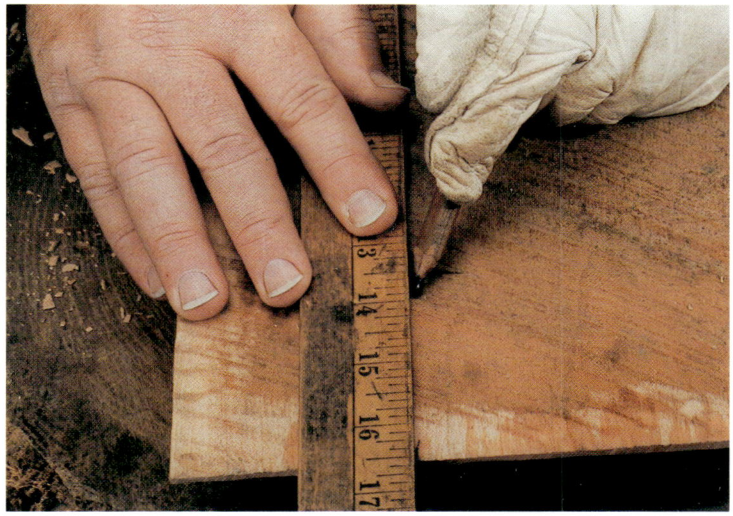

from each of the 3" marks.

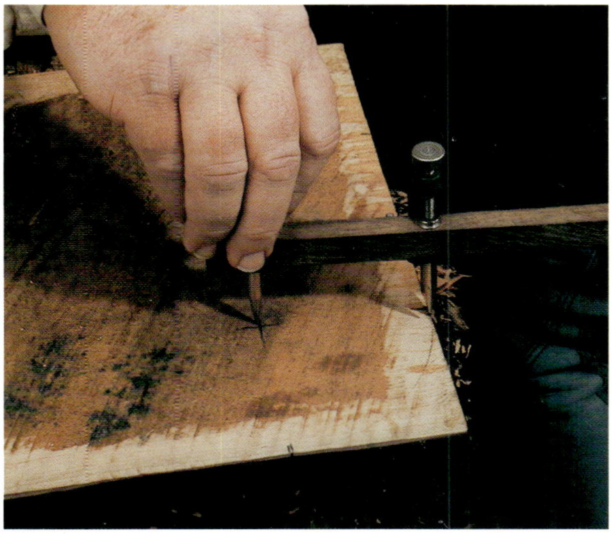

With the point of the compass at each of the X's, strike an arc at each corner.

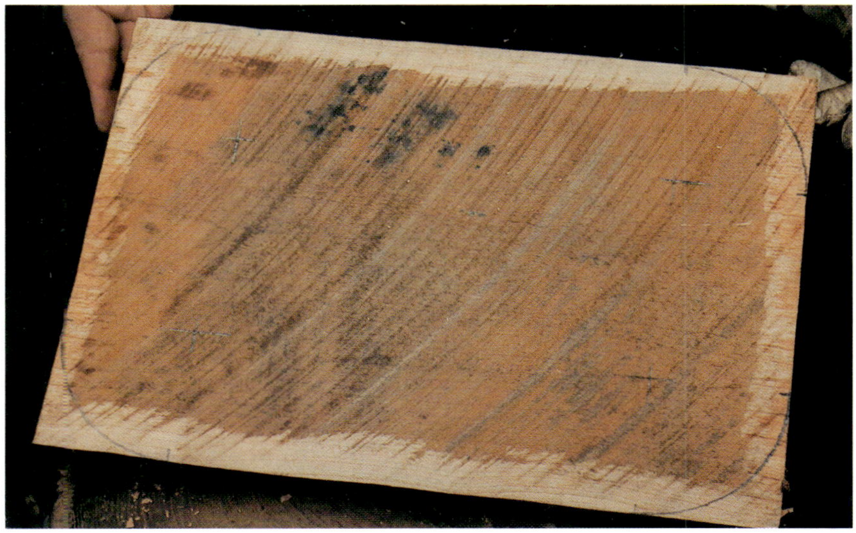

The corners are marked.

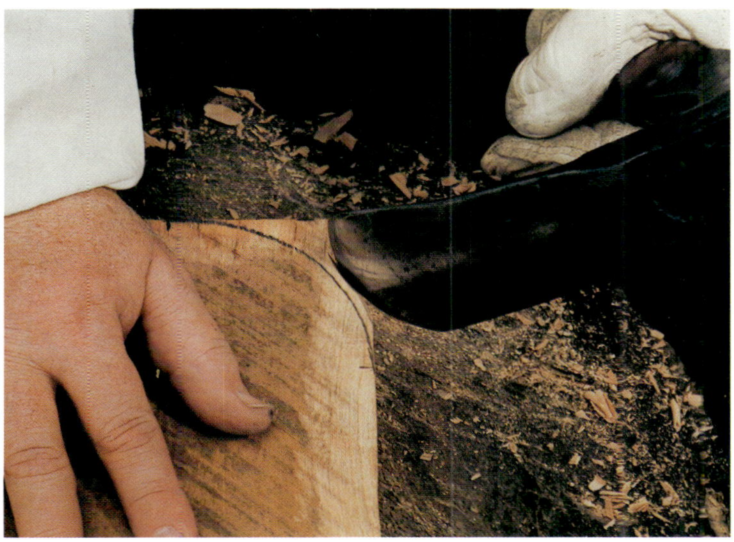

Rough trim the corners to about 1/4" from the arc. You must go with the grain.

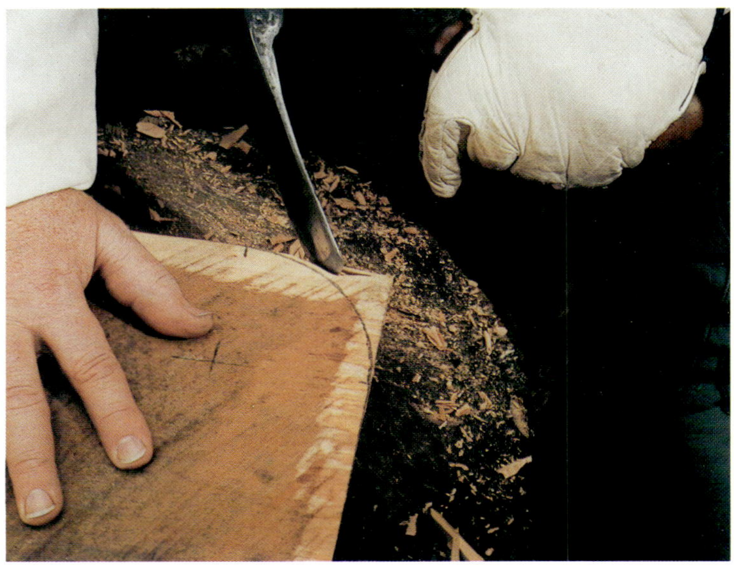

This will require you to go backhand at two of the corners.

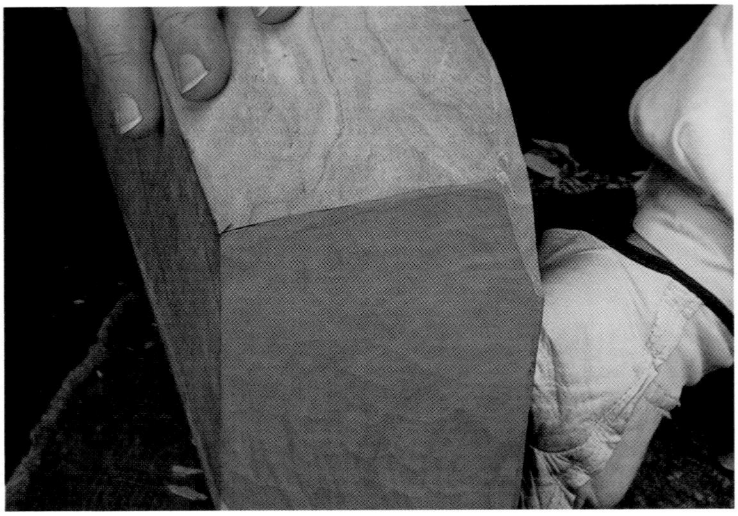

The result. Now turn the slab over to begin rounding the corners.

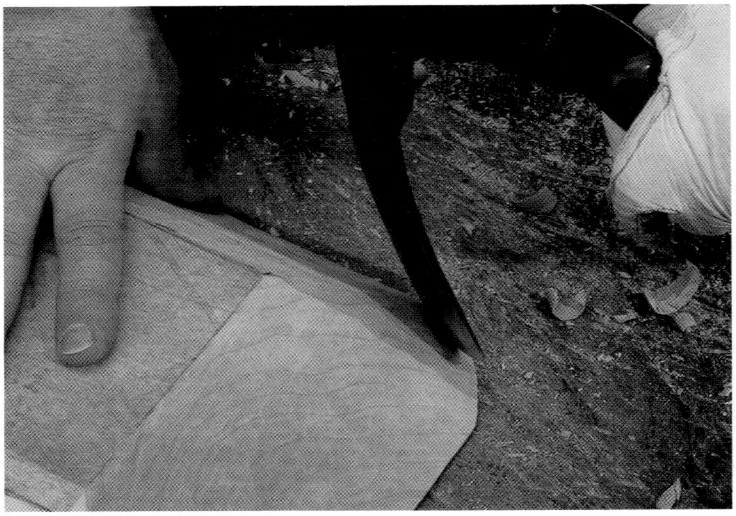

Chop at an angle with the grain.

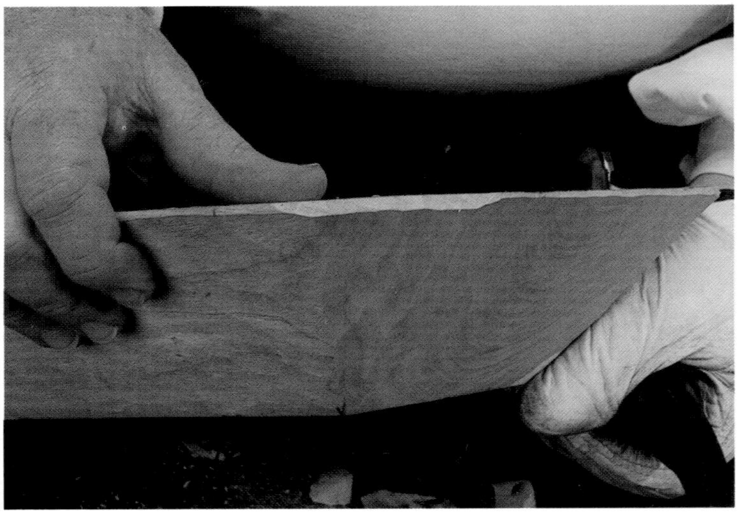

Progress.

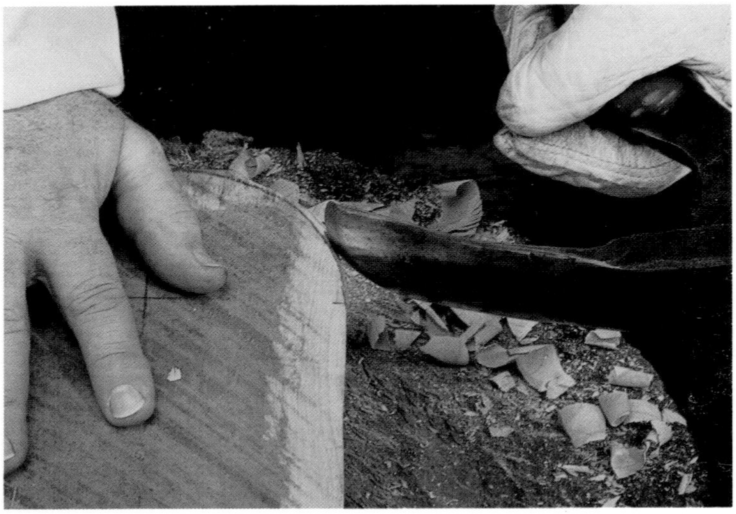

Turn the slab over and take the corners closer to the line. Again, go with the grain.

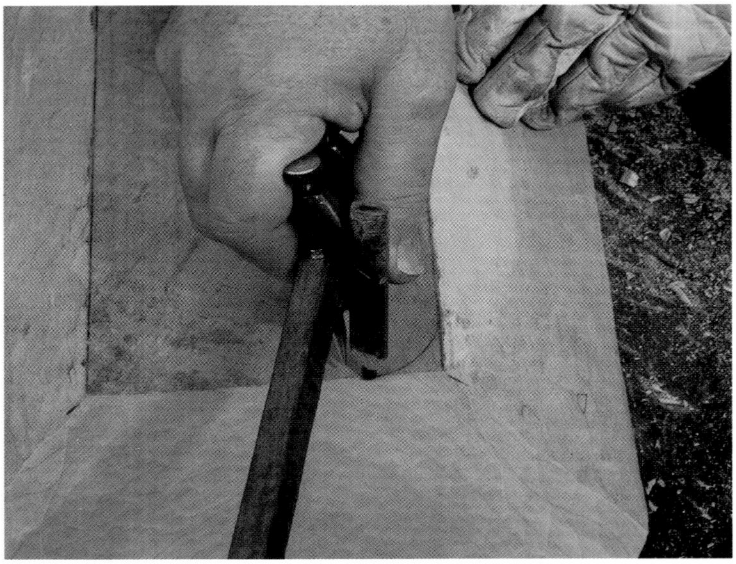

On the bottom of the bowl follow the same procedures to strike an arc with a 1 1/4" radius at each corner.

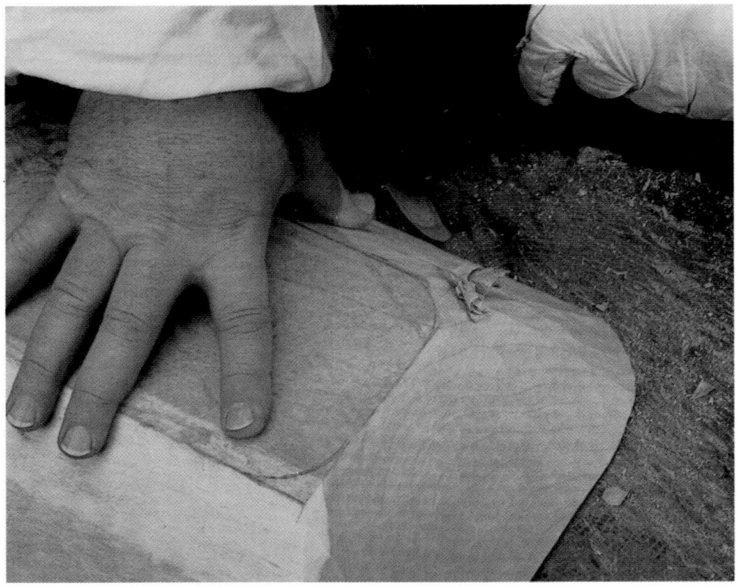

Rough cut back to the mark.

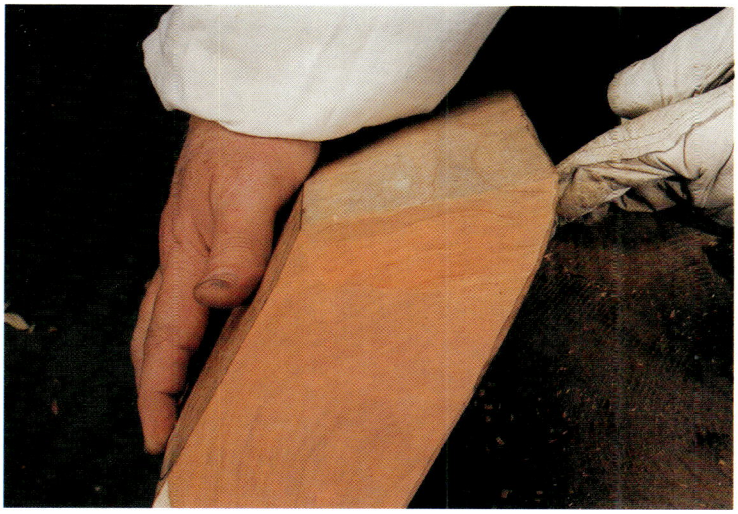

bringing the corner down so that the edge is thinner.

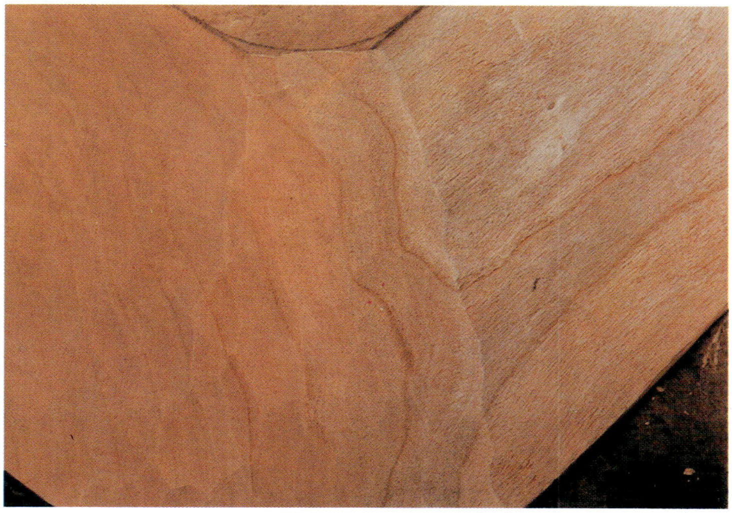

The corner roughed in.

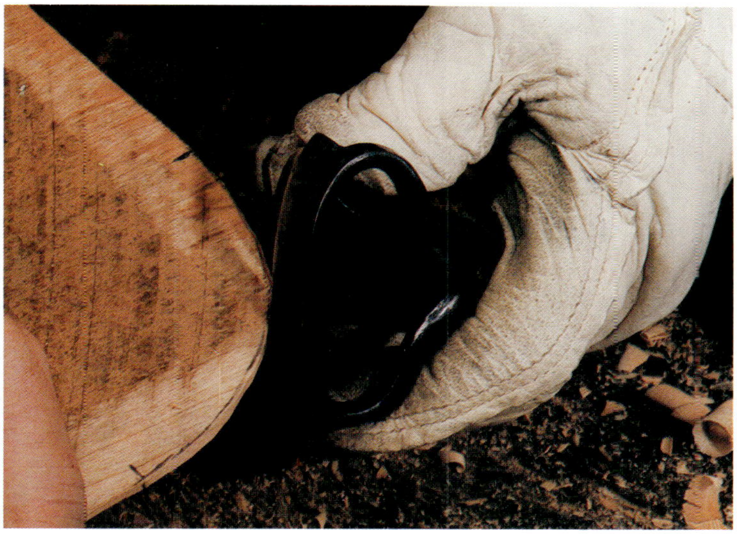

Going with the grain, plane or rasp the corner to make it smooth.

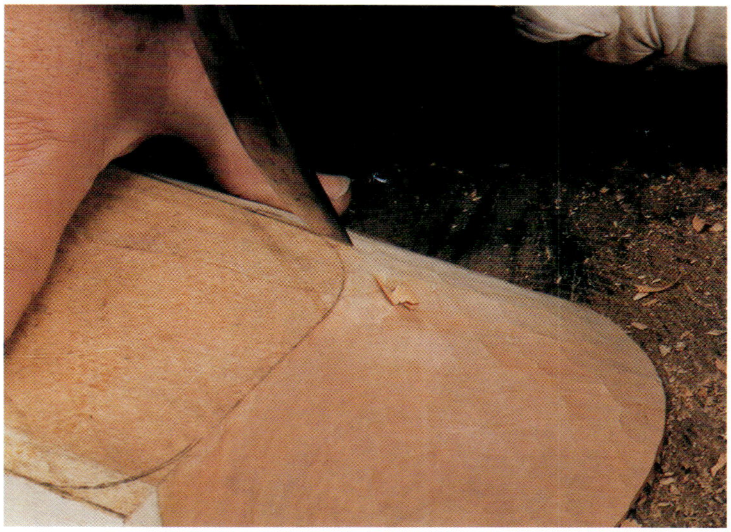

Start at the line and use short strokes to smooth the corner.

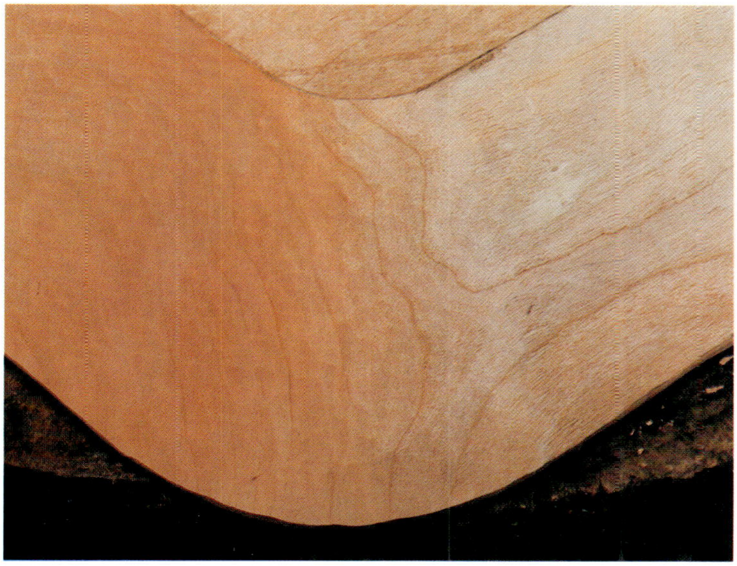

One corner finished. Repeat the process on the other three, always remembering to work with the grain.

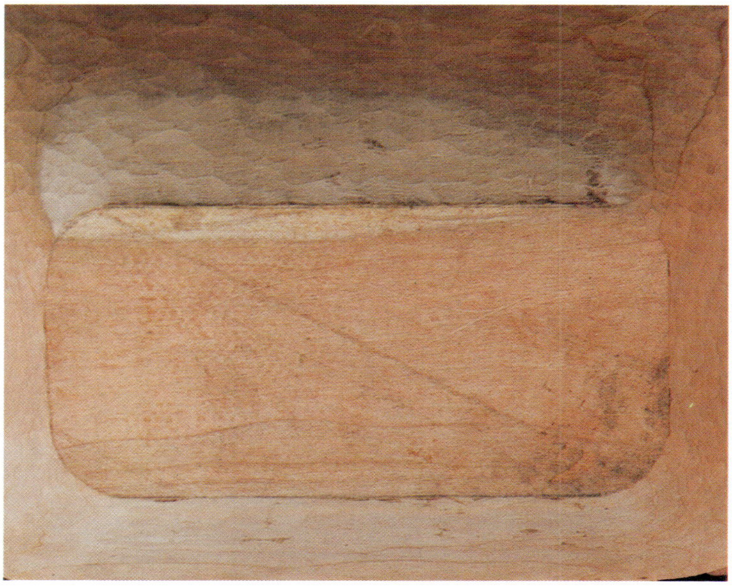

The corners finished.

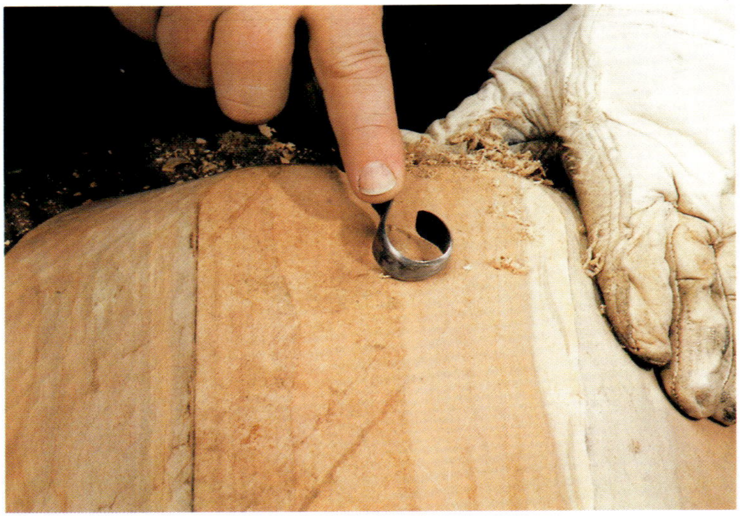

Clean up the bottom with a scraper or scorp.

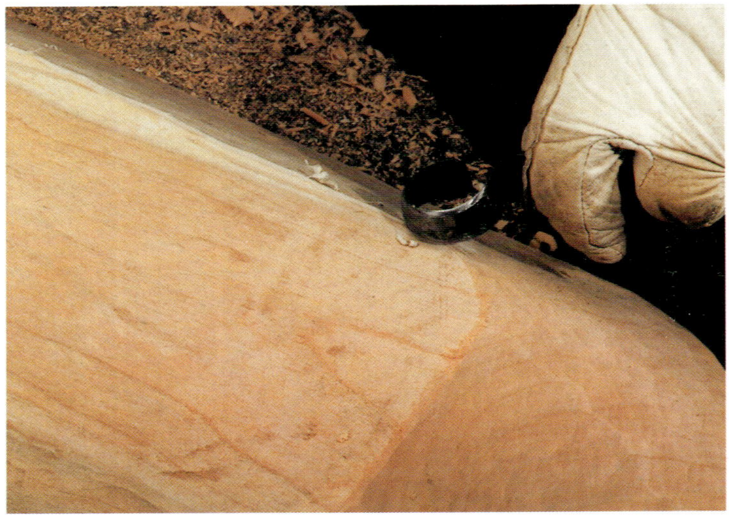

Slightly bevel the edge to prevent chipping.

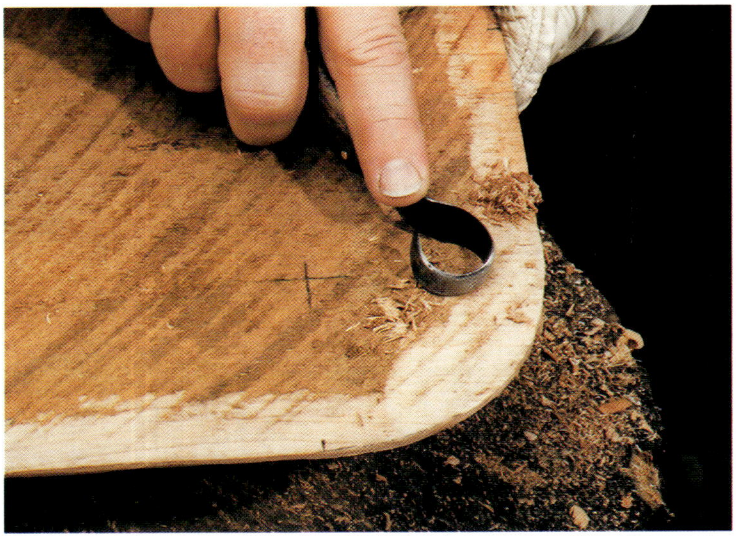

Clean around the top, about 3/4" from the edge. The thickness of the finished wall should be somewhere between 1/2" and 3/4".

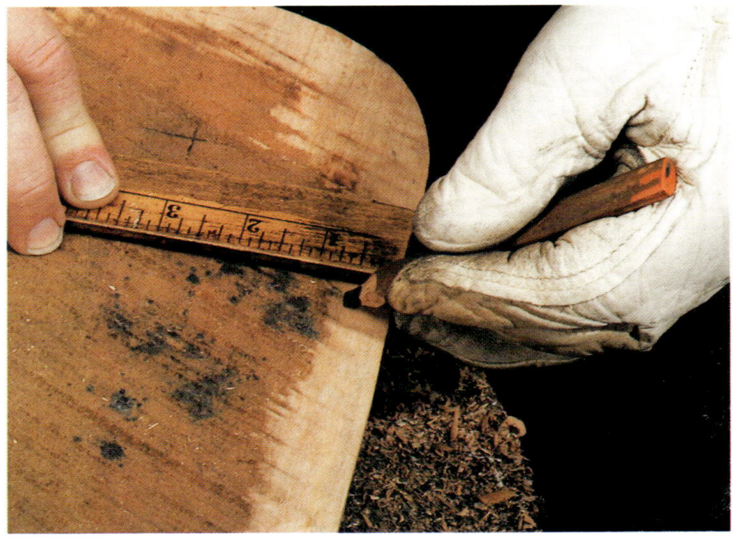

Mark the thickness of the wall all around...

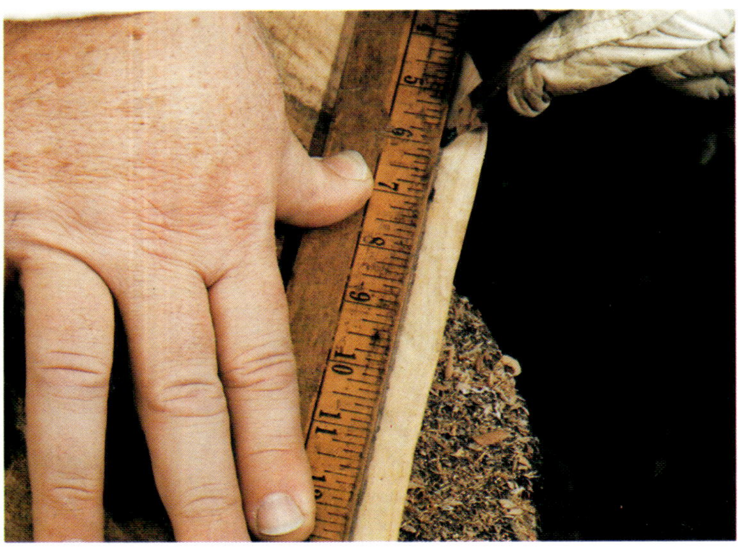

and draw the line along the sides and the ends.

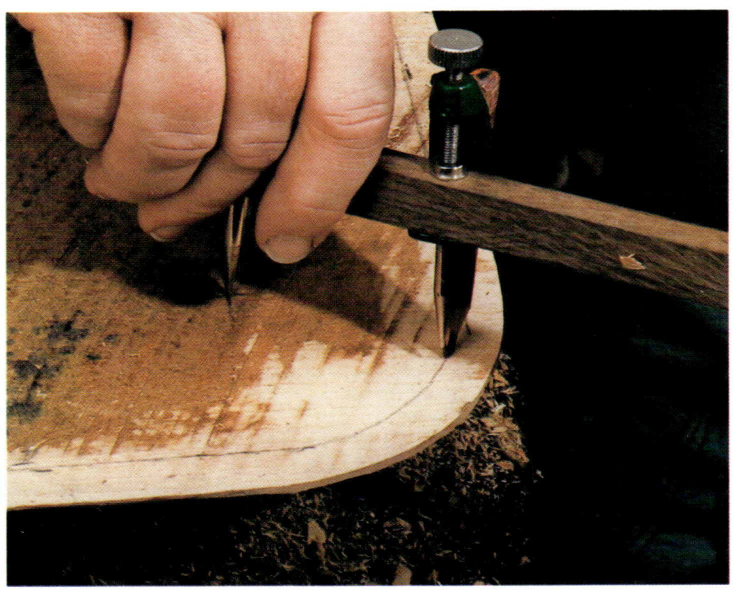

Using the same pivot point as before, set your compass to draw the arc of the corner at the correct thickness.

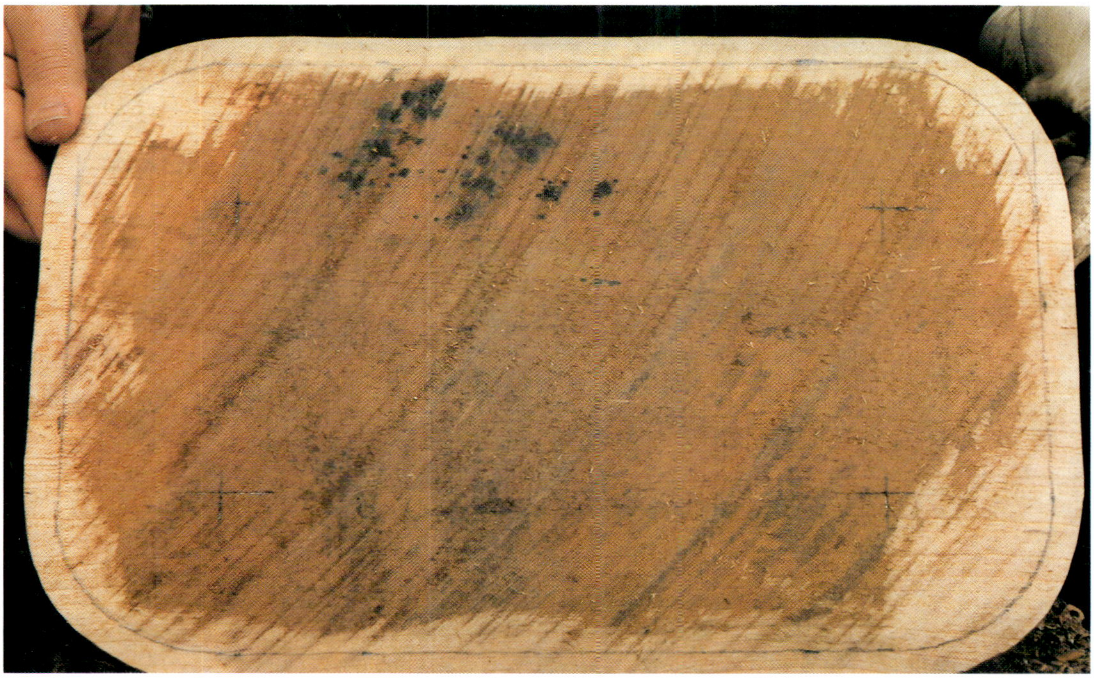

The wall thickness drawn.

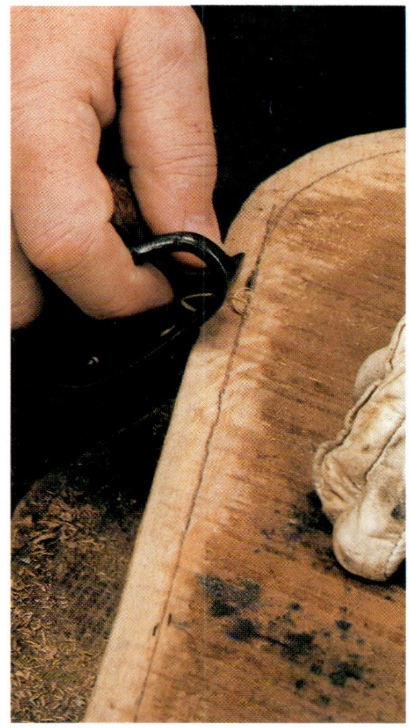

Slightly bevel the edge of the bowl. This gives you an edge that won't cut you (it has happened!) and is less apt to be chipped while you're working.

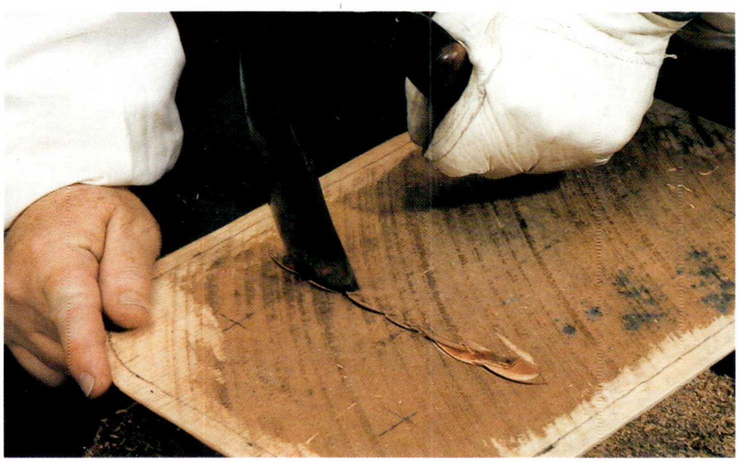

I begin removing the inside of the bowl by cutting across the grain at both ends. This creates a relief cut by cutting the fibers. First make a series of cuts straight across the grain.

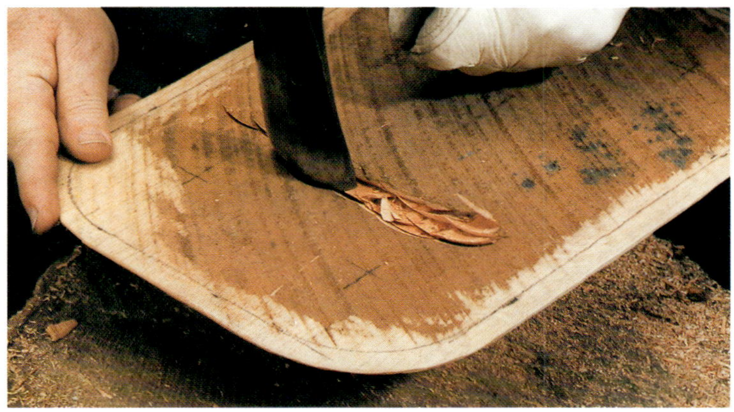

Then move about a half inch out and cut across again, creating a trough.

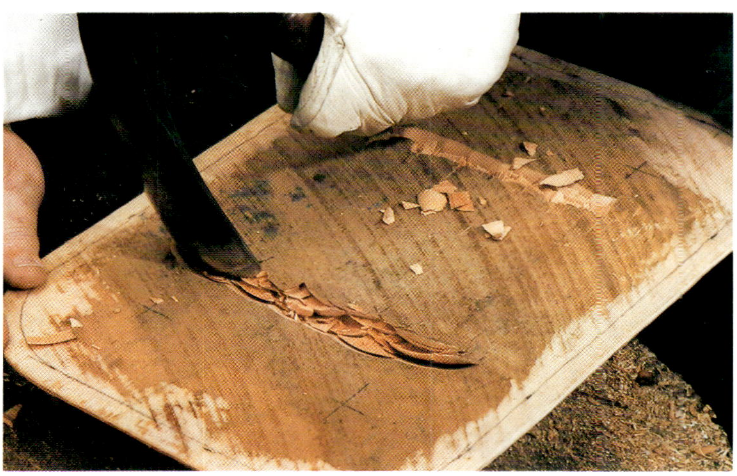

Repeat at the other end.

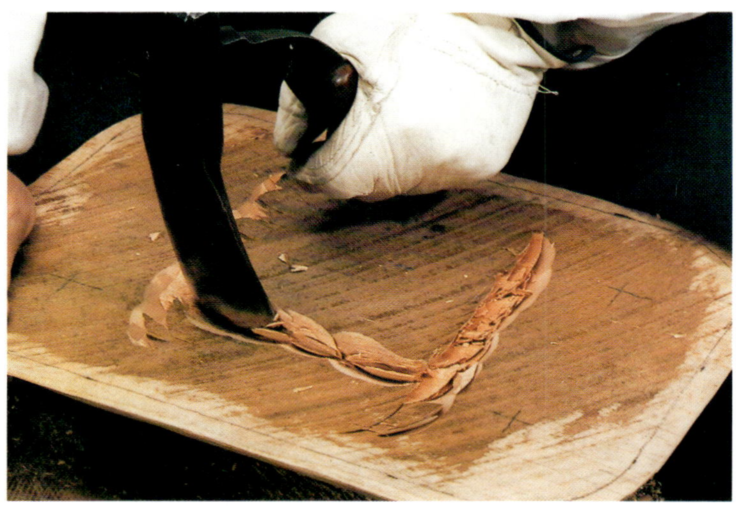

To remove the center you work sideways, making one cut...

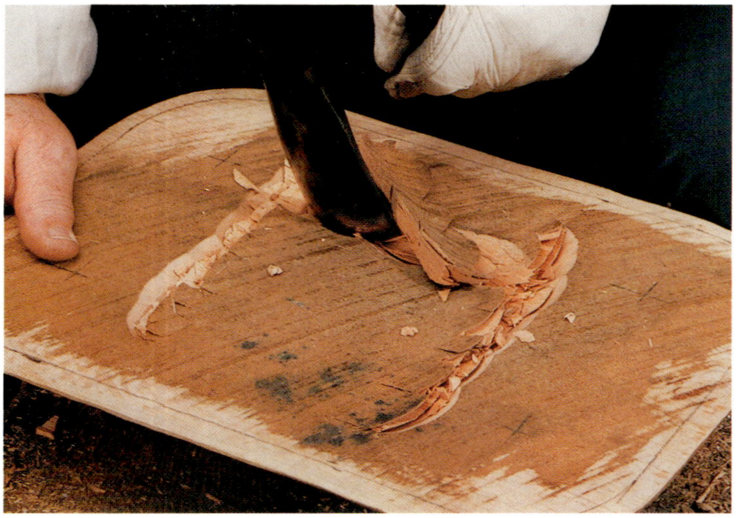

then coming back to it. At this point it is truly a bowl because it can hold something.

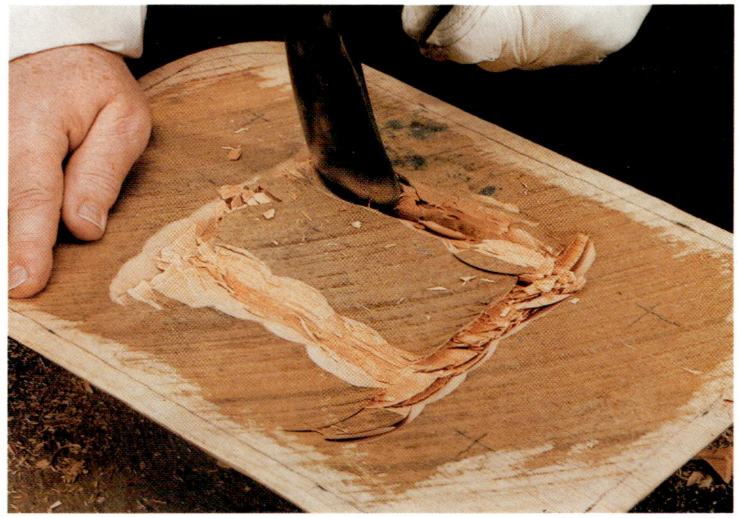

Do the same thing on the other side...

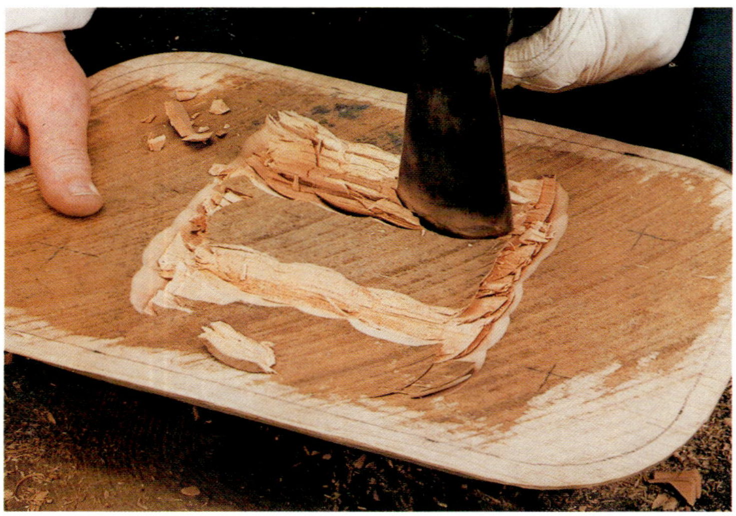

then work your way across, about a half inch at a time.

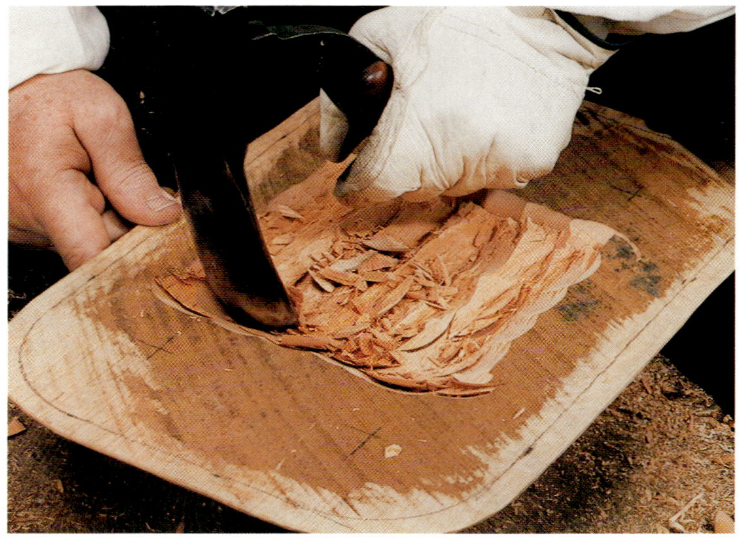

When you've finished going across, go back and deepen your endgrain cuts at both ends.

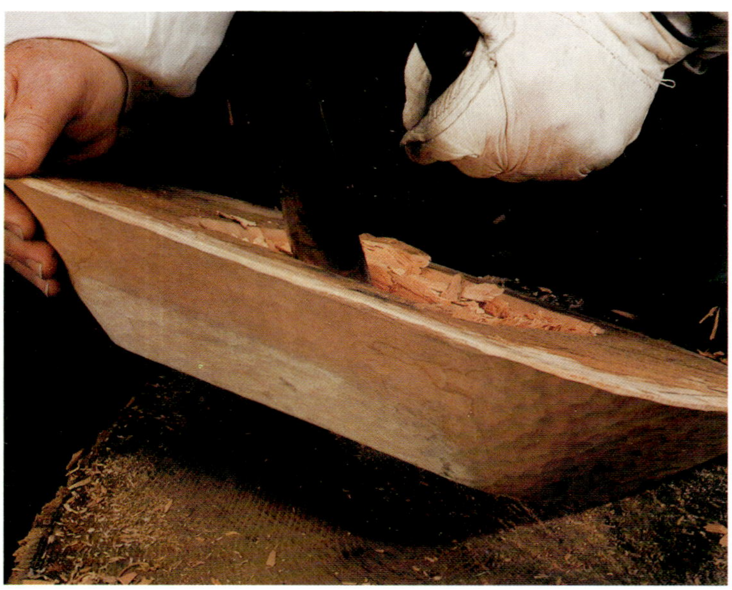

Tilt the bowl up and gently clean it up going across the grain. My knee acts to brace the bowl to keep it from sliding.

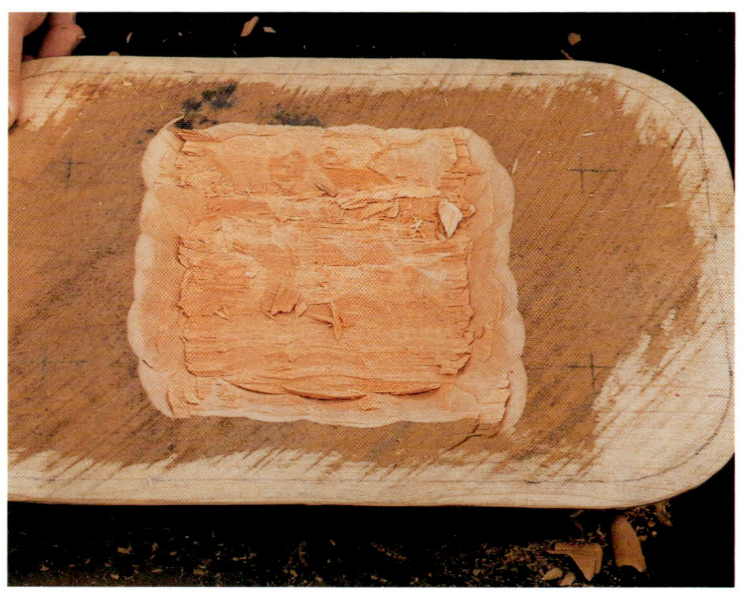

This takes us to this point.

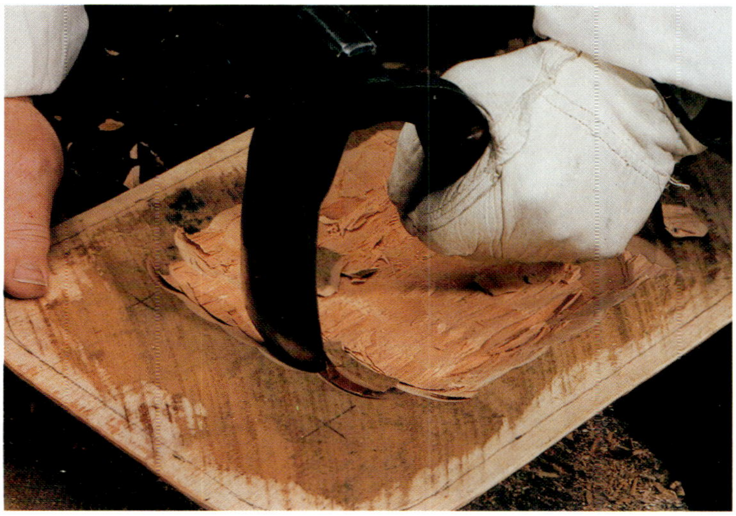

Move about a half inch toward each end and cut across the grain again.

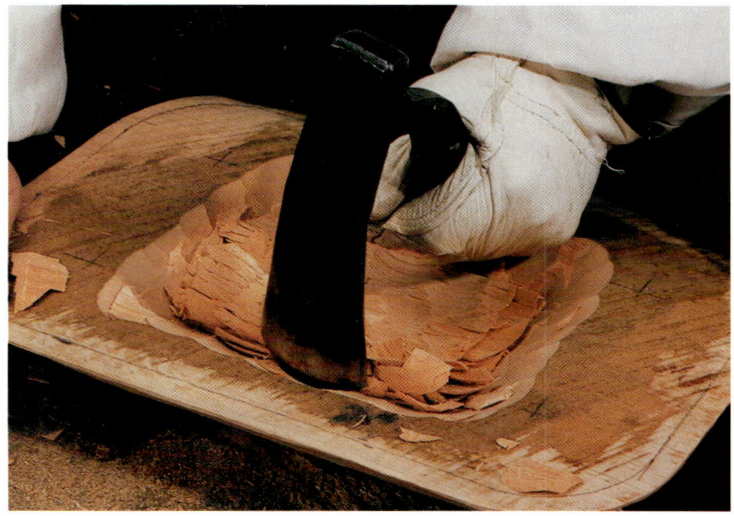

Come around to the sides, but don't go a half inch. There simply isn't as much to remove on the sides.

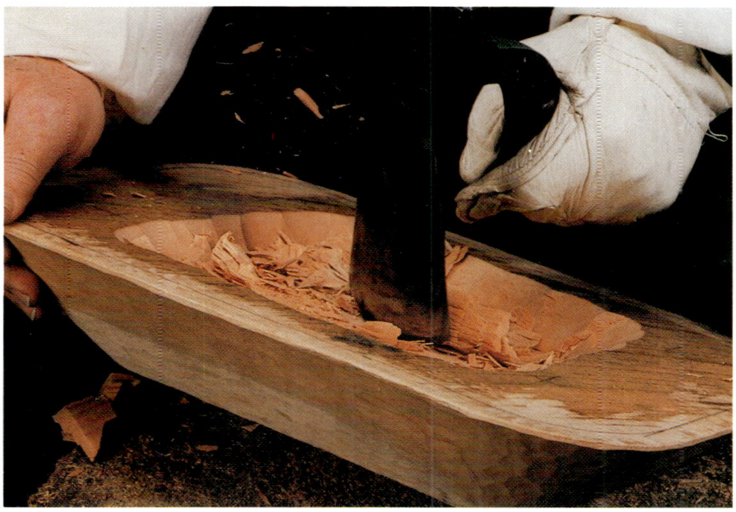

Again, tilt the bowl up and deepen the bottom.

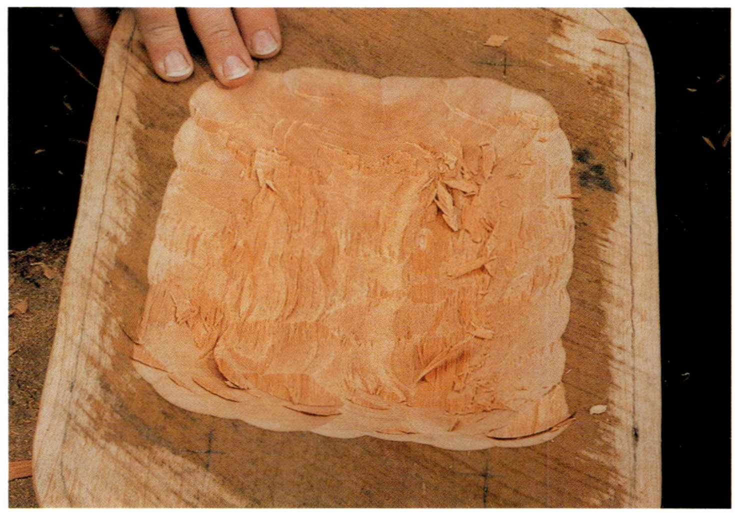

This process takes us deeper and closer to the edge in each direction.

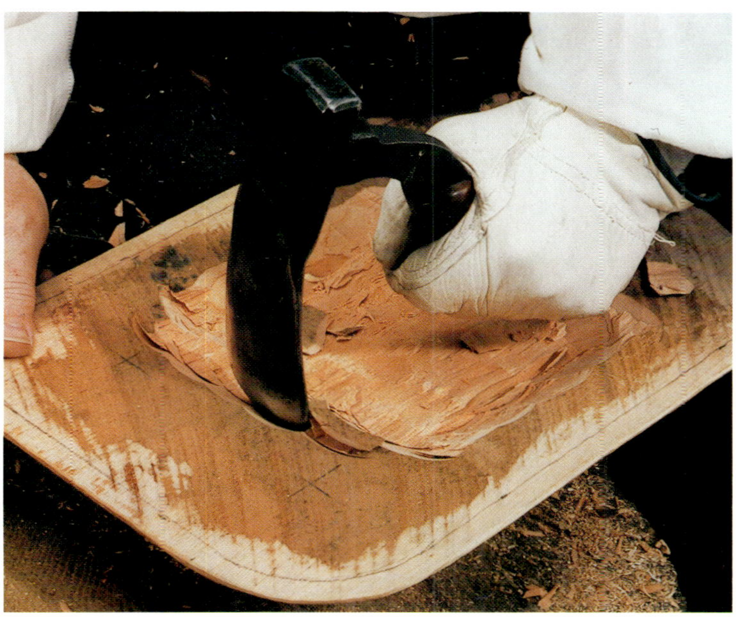

Continue the process, going a half inch toward the ends...

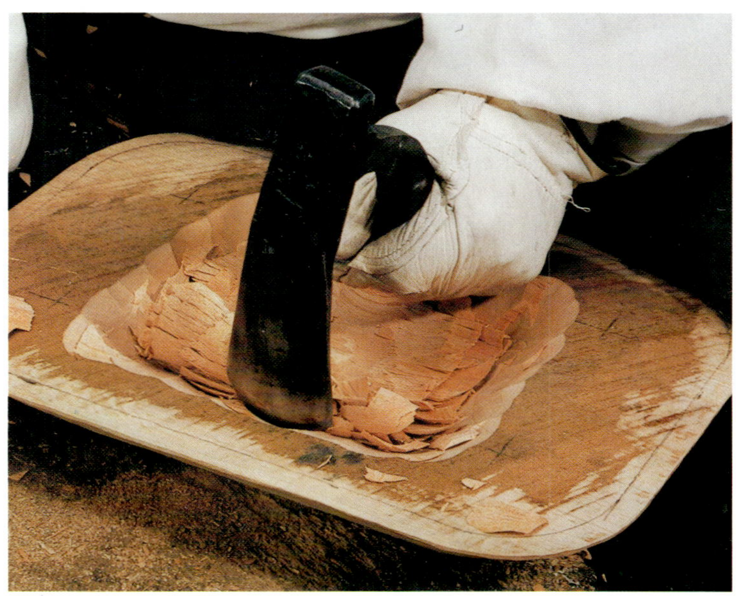

and somewhat less along the sides.

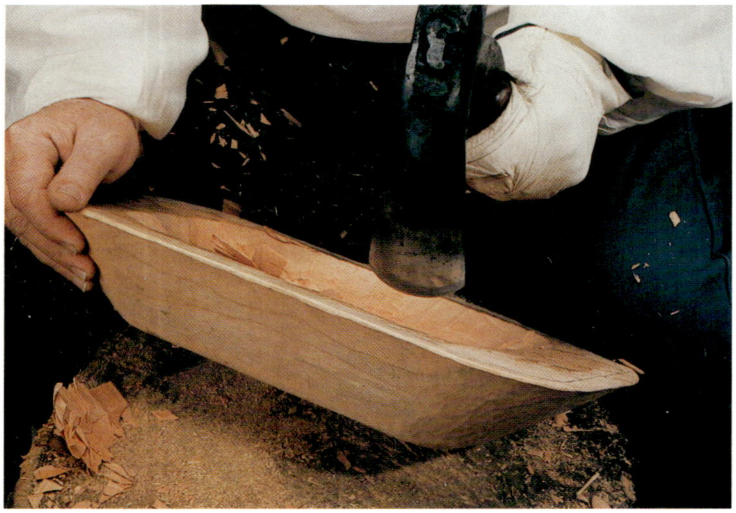

Deepen the bottom.

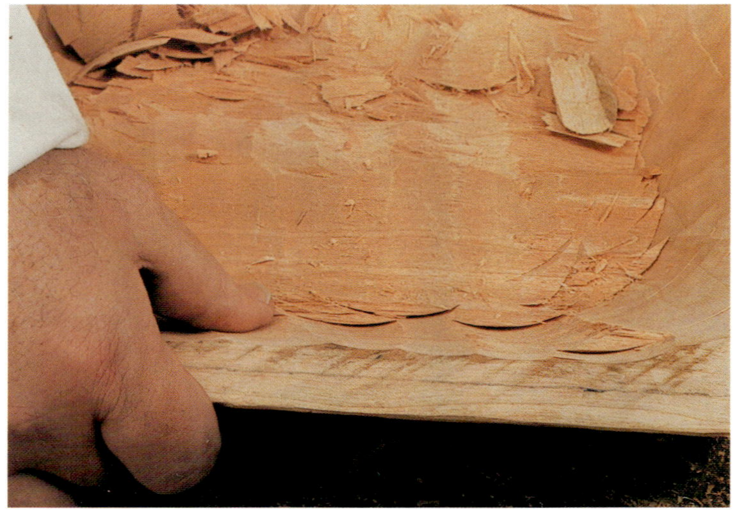

Check by eye and by touch to be sure the inside and the outside of the bowl are parallel, follow the same plane, and have uniform thickness.

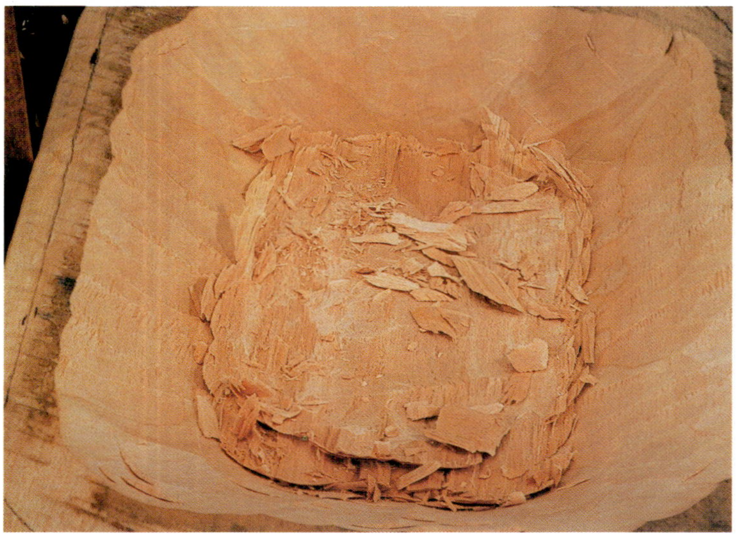

You want the bowl to have a bottom that is about 1/2" thick. To measure the thickness, measure the height of the bowl as it stands on the table.

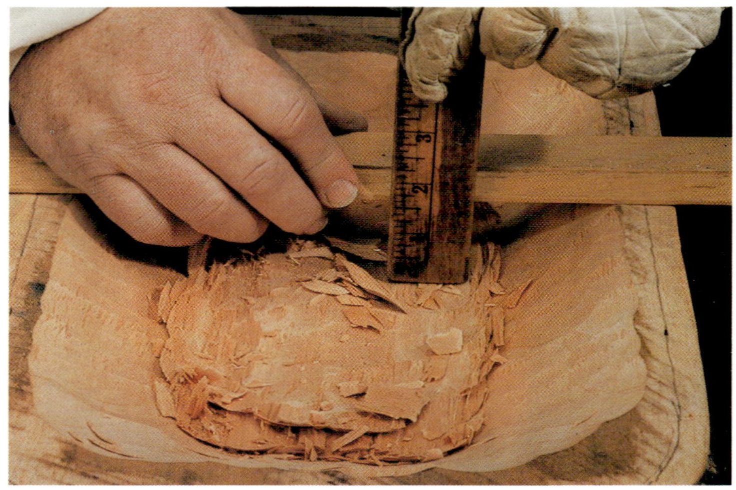

Lay a straight edge across the top, and measure the depth. Since the bowl is 3" tall, the final depth should measure 2 1/2".

Take the center to about 1/2" from the line all around.

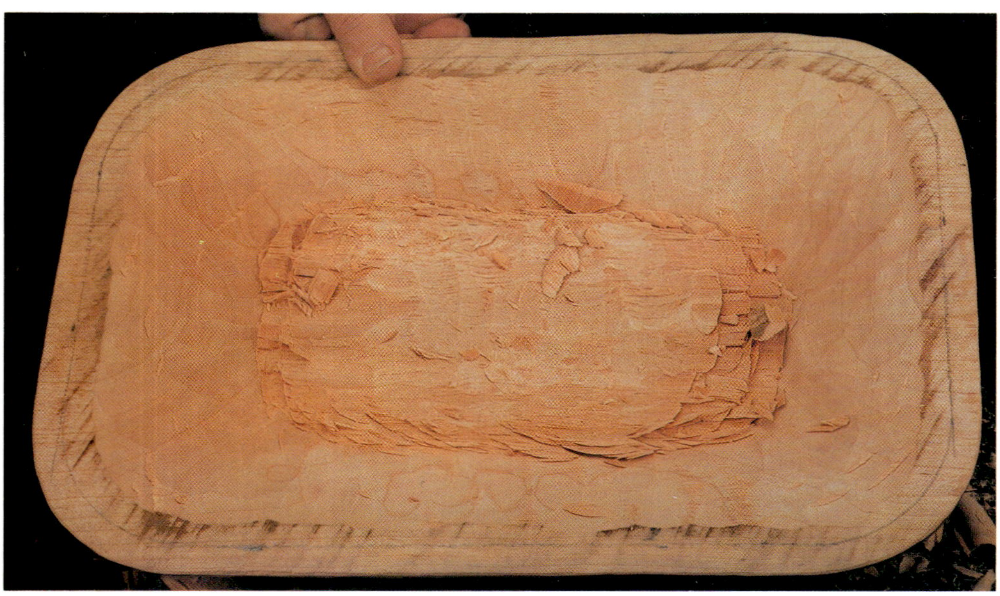

Clean across the bottom...

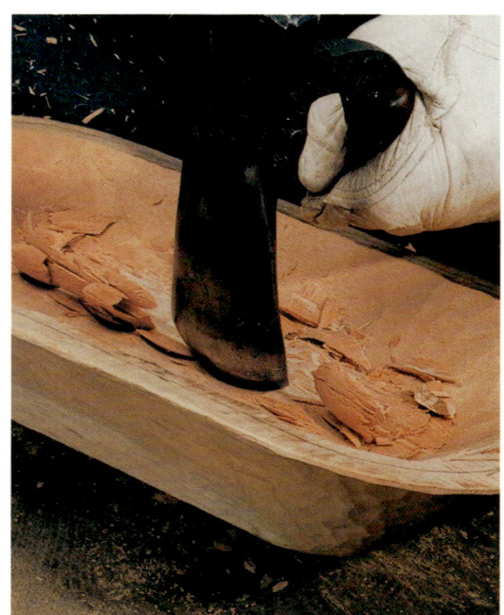

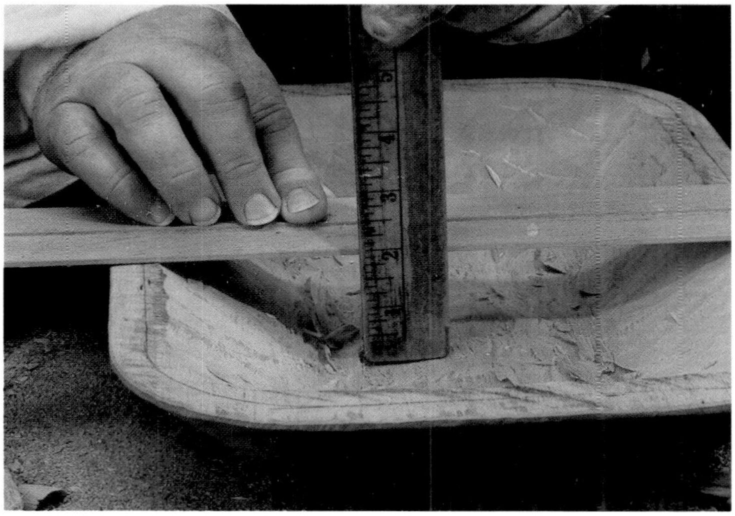

and check the depth again.

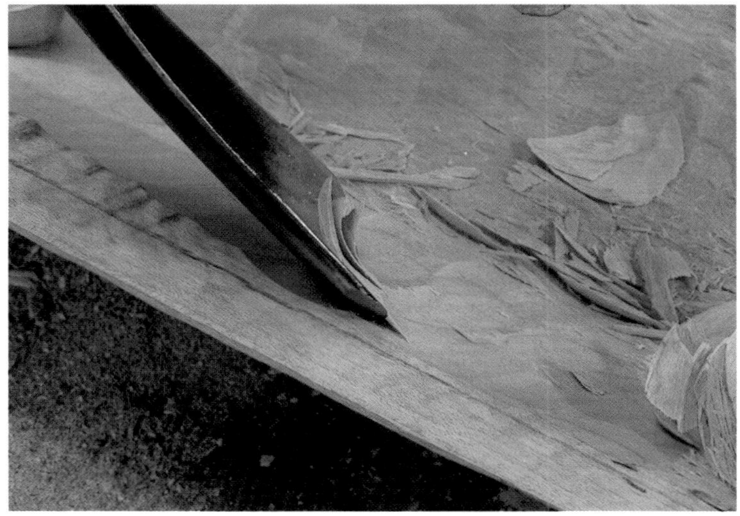

Going with the grain, reduce the top inch or so of the sides, taking it to the line.

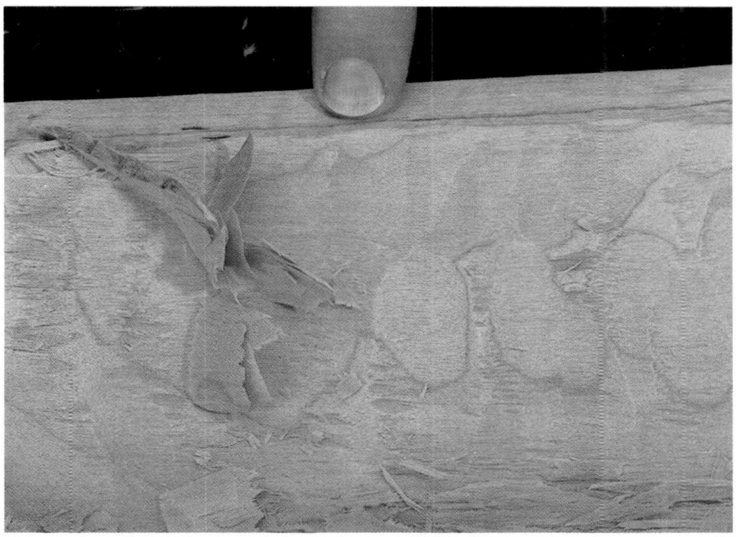

When done correctly it should leave you a nice smooth surface.

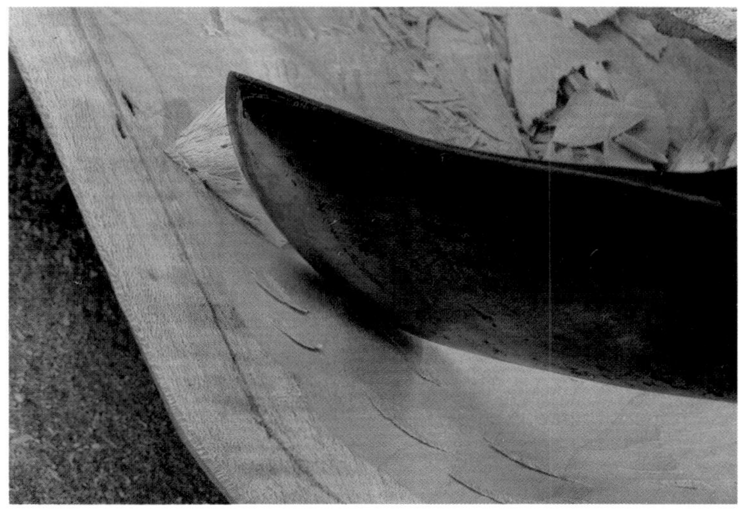

If you go against the grain...

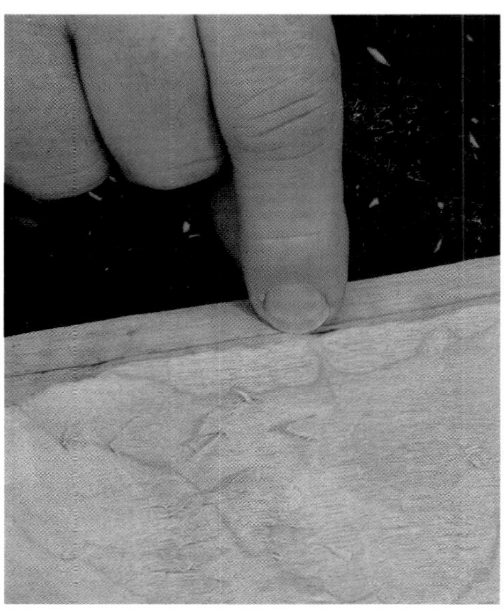

you are likely to end up with tears like this.

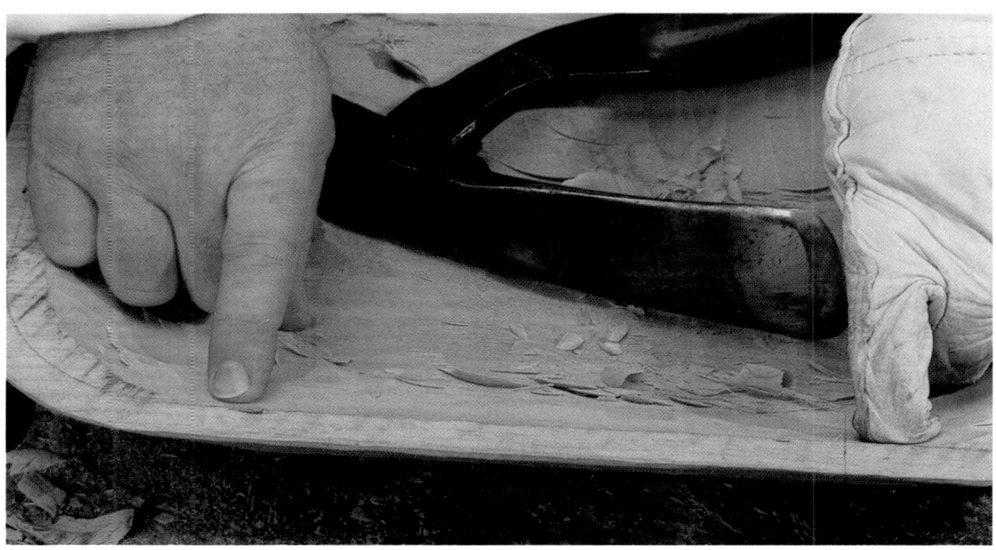

Thin the top of both long sides to the points where the corner arcs begin.

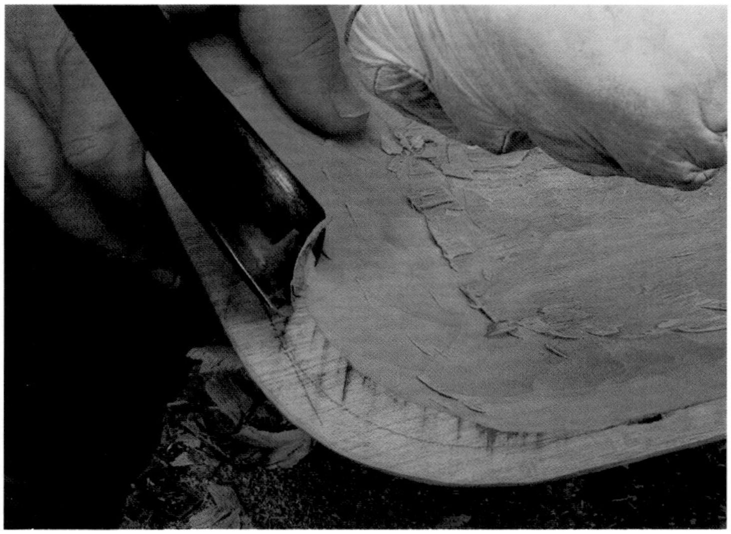

To continue thinning the top, work from the center of the end back to the corner. This assures that you will be going with the grain as you make the turn. The adze should be angled slightly as you go across the end grain.

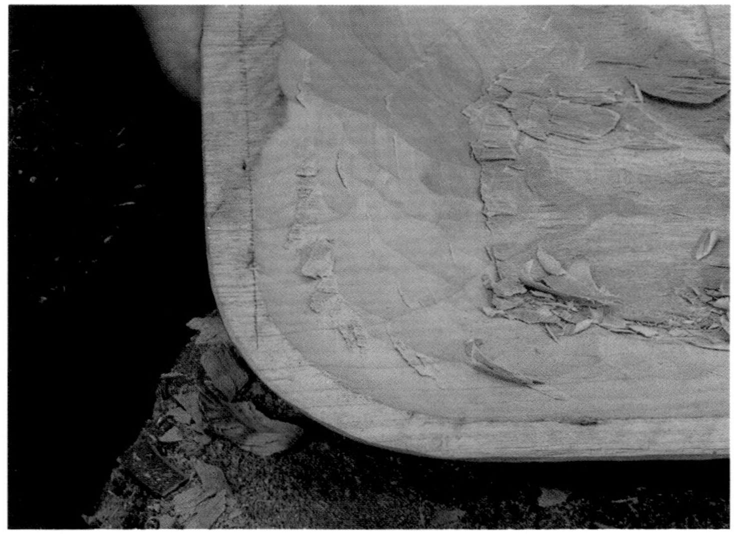

That will take you to this position.

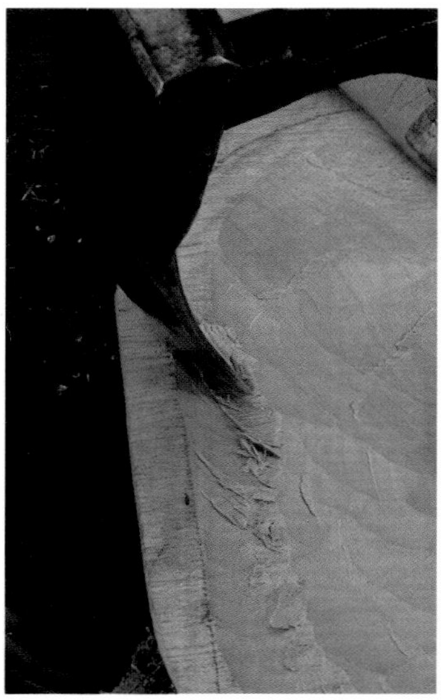

Repeat the process in each corner.

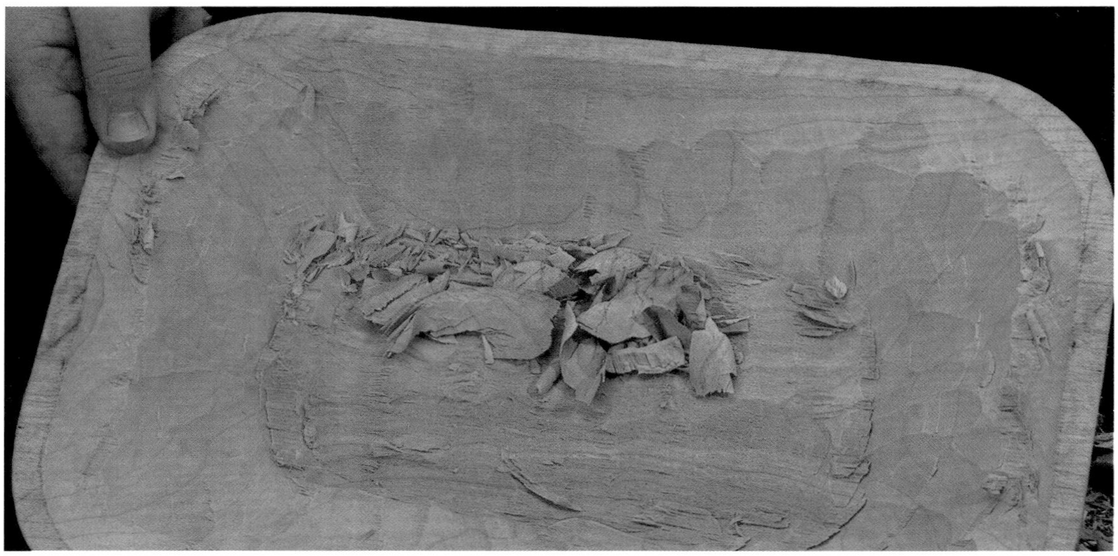

The top edge reduced to the thickness we want.

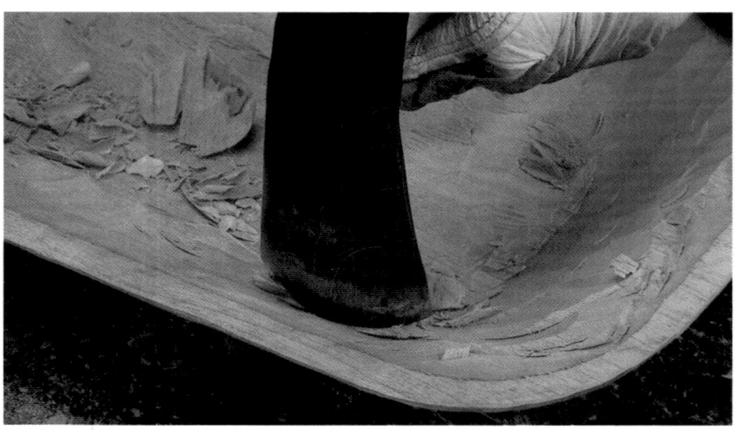

Reduce the sides, starting below the area you just thinned and working down to the bottom. The bottom is nearly as deep as it should be, so don't go too deeply.

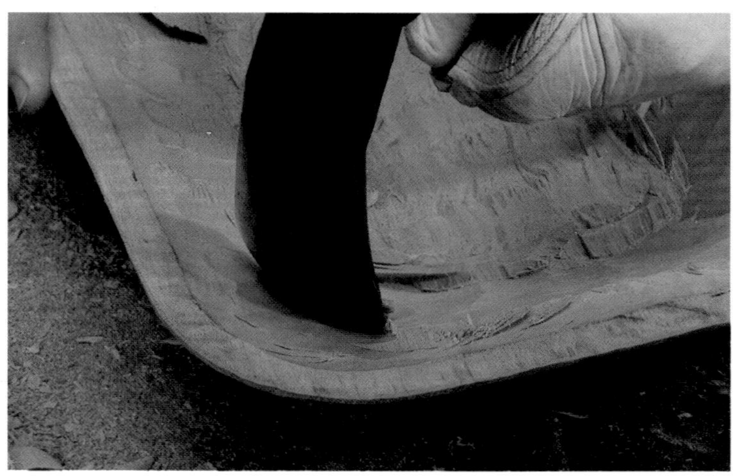

Continue all the way around the bowl.

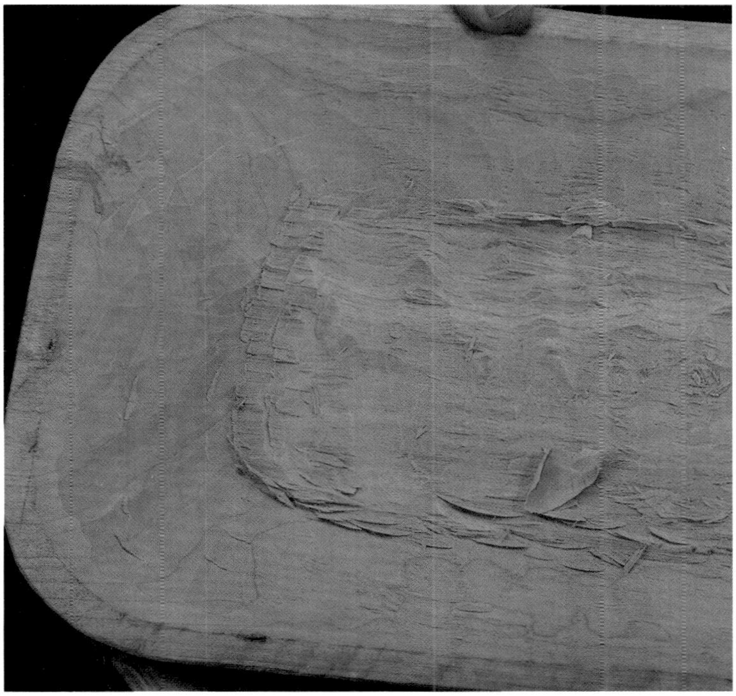

This takes it to the next-to-finish cut.

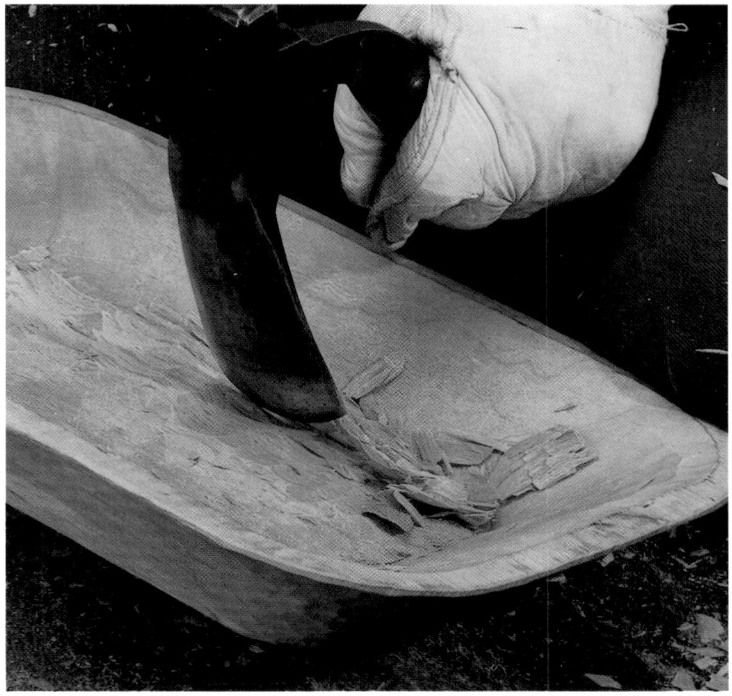

Straighten up the bottom. It was thicker at the ends than in the middle.

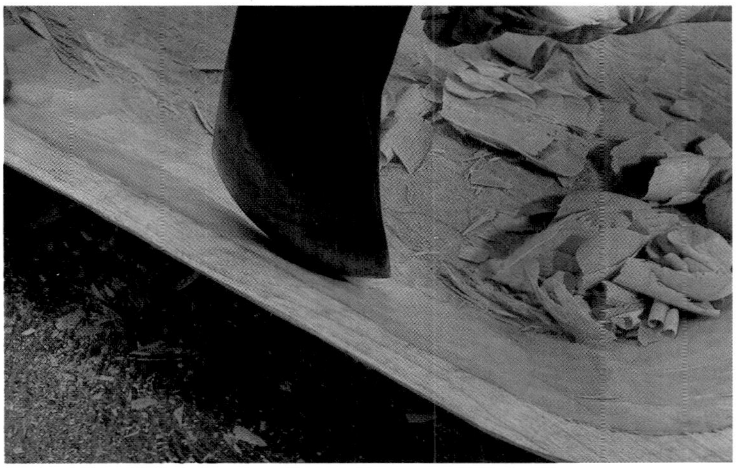

The finishing strokes start at the top...

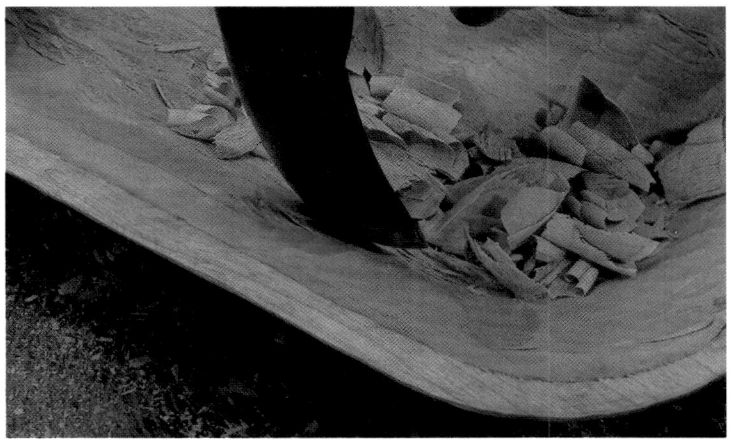

and go to the bottom. This up-to-down pattern is followed around the inside of the bowl.

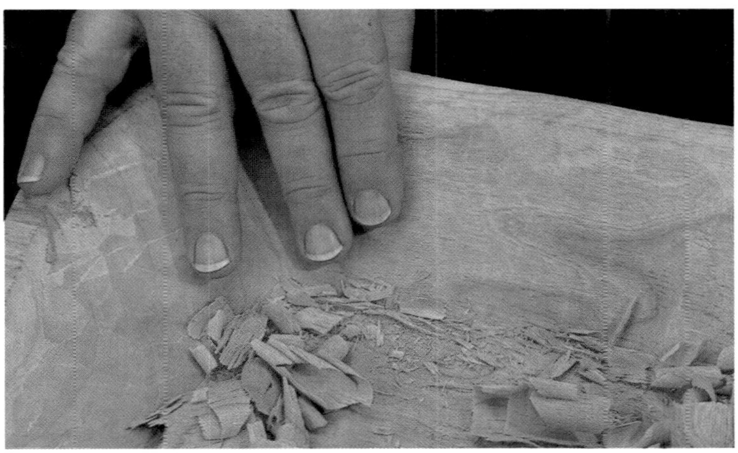

Stop frequently to check the thickness of the wall for evenness.

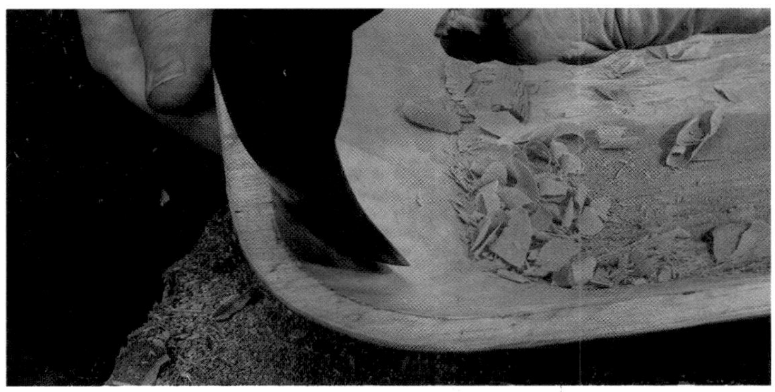

The end grains are much more fragile than the side grains. Be sure the bowl is secure against the work surface and cut at an angle. The old bowl makers left the ends a little thicker. This helps equalize the drying times with the sides, and is an added protection against warping and checking.

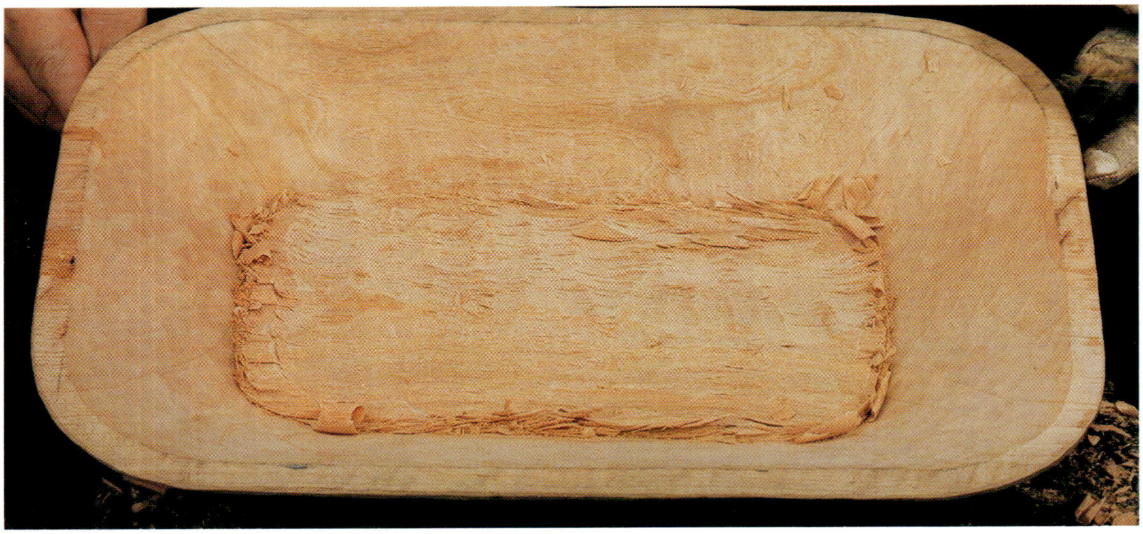

The walls finished.

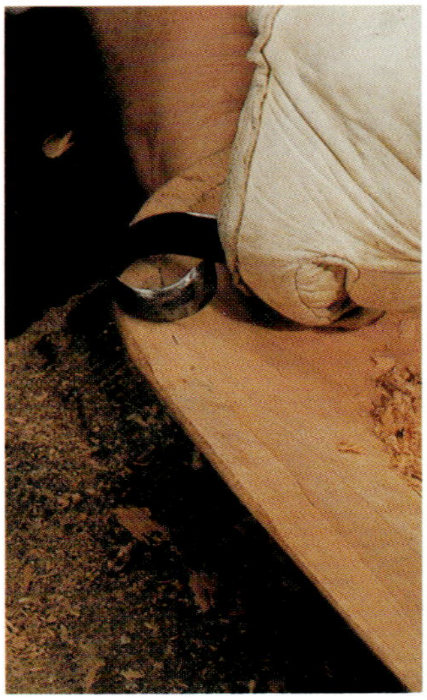

Clean up the edge with the scorp, removing pencil marks and any other flaws.

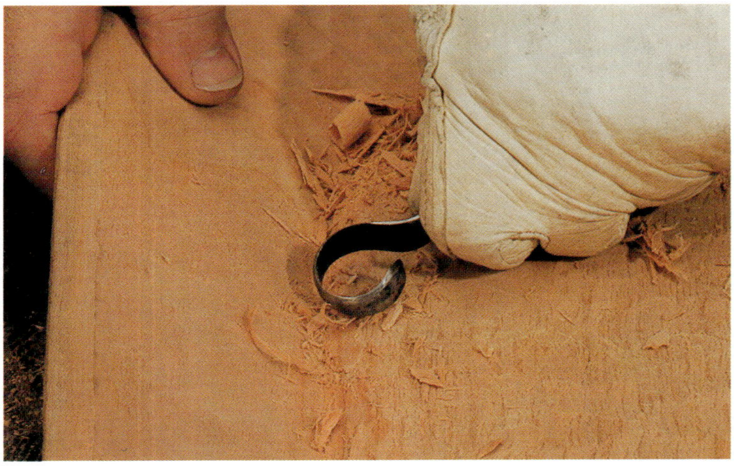

Use the scorp to clean up the edges, where the bottom meets the sides.

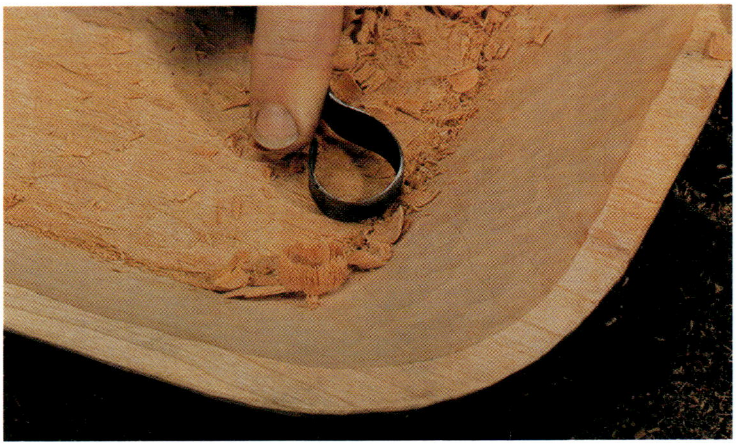

Go across the grain at the ends and do the same thing.

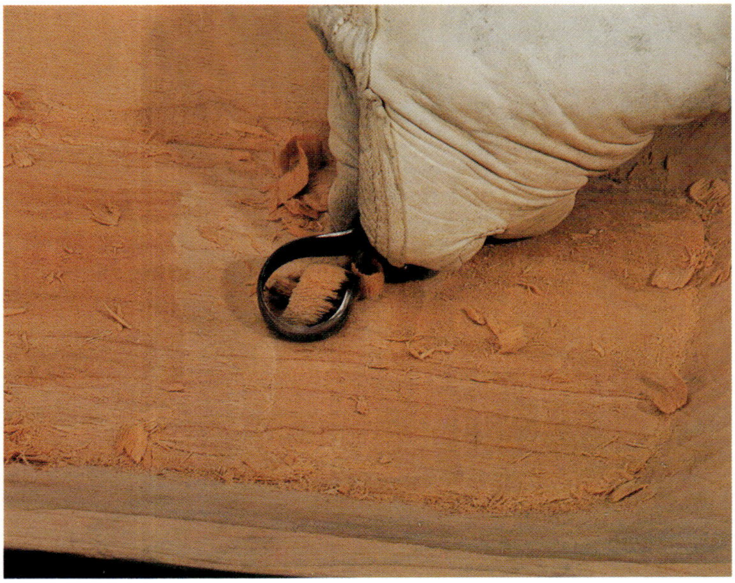

Continue going cross grain going down the length of the bottom. This removes heavy stock.

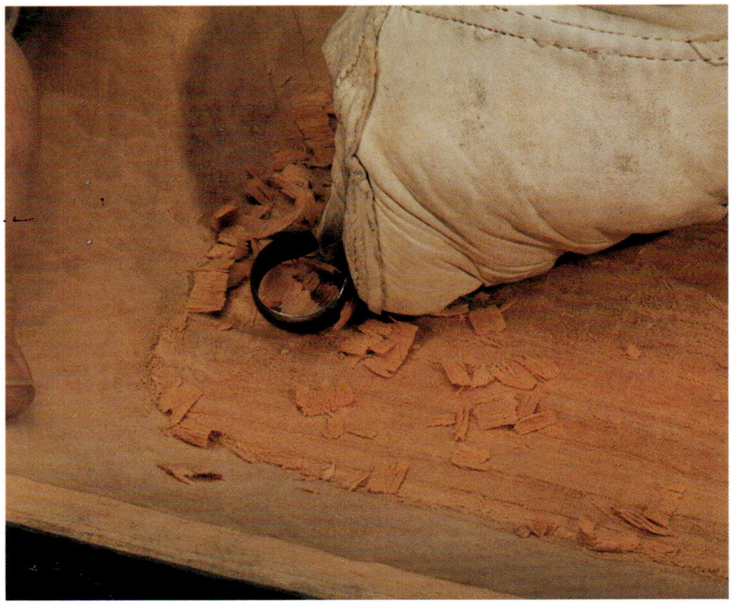

Go over the bottom cross grain two or three times.

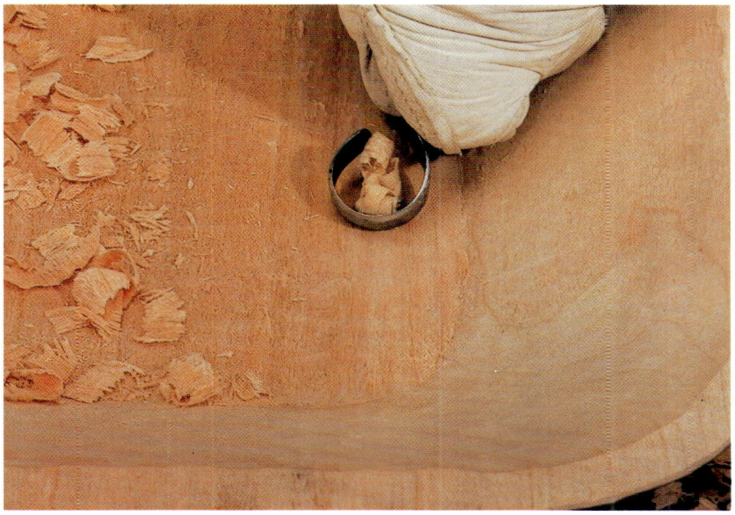

You may need to go into the corners with the grain. This will smooth it out enough so...

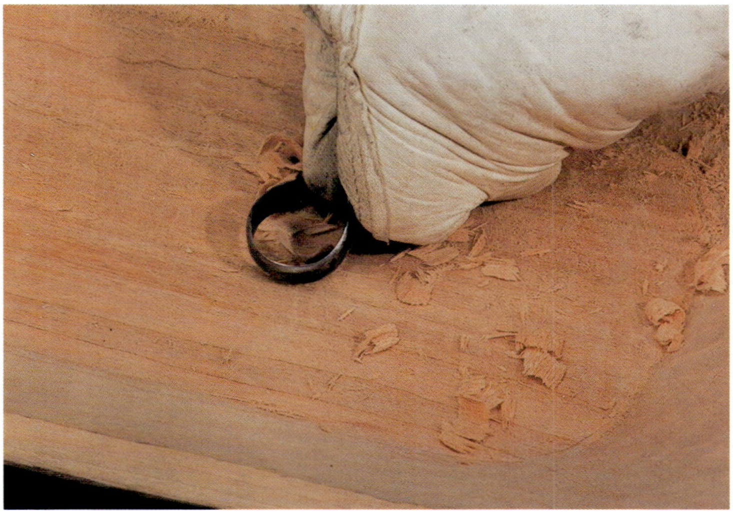

you can return to going across the grain.

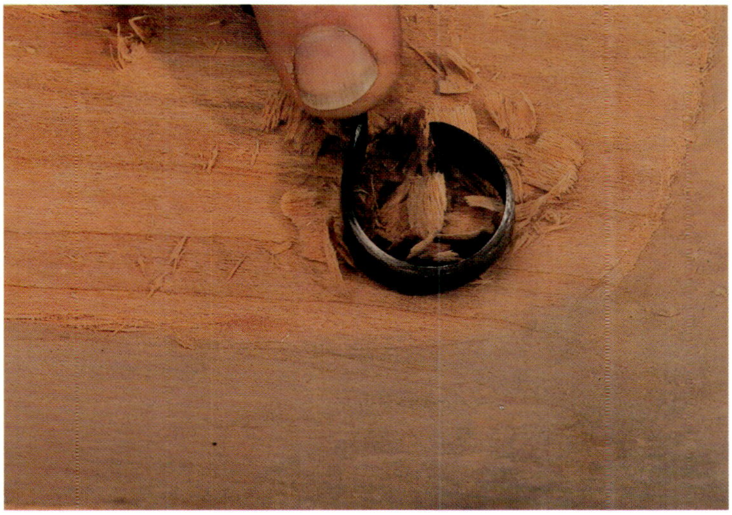

In thinning the sides the adze goes deeper into the bottom in some places than in others. I'll continue working the bottom across the grain until I have worked these out.

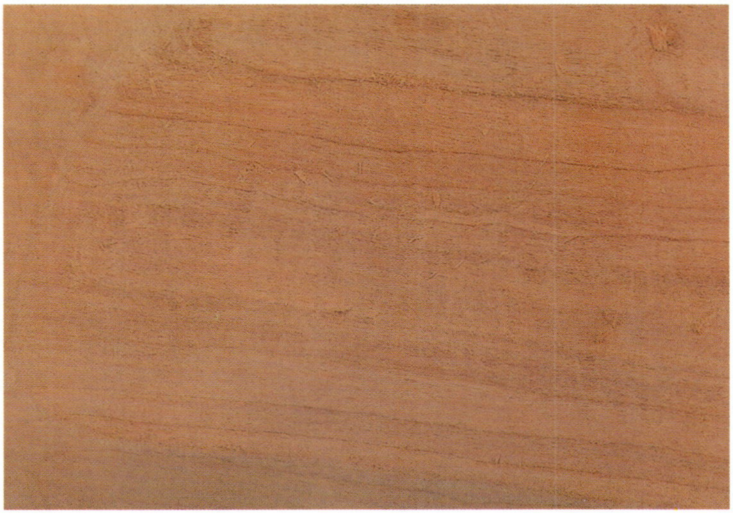

The cross grain scorping will give a smooth, flat surface, with a soft transition from the sides to the bottom.

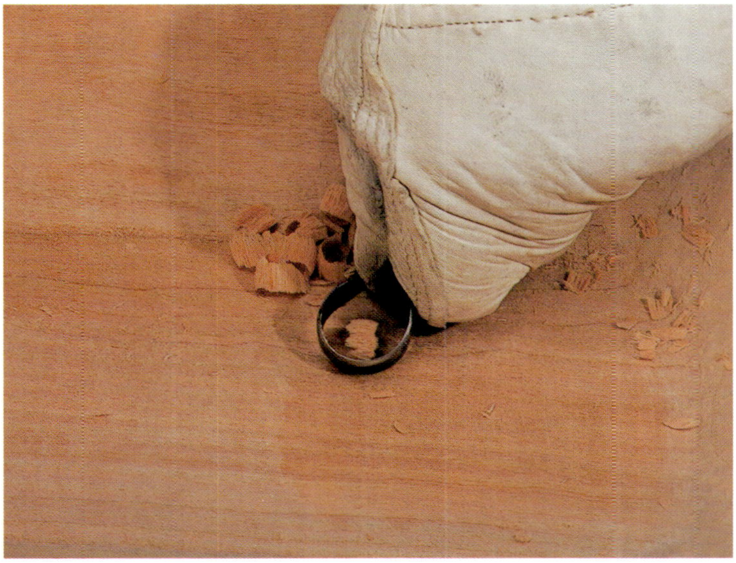

Work across the grain all around the bottom...

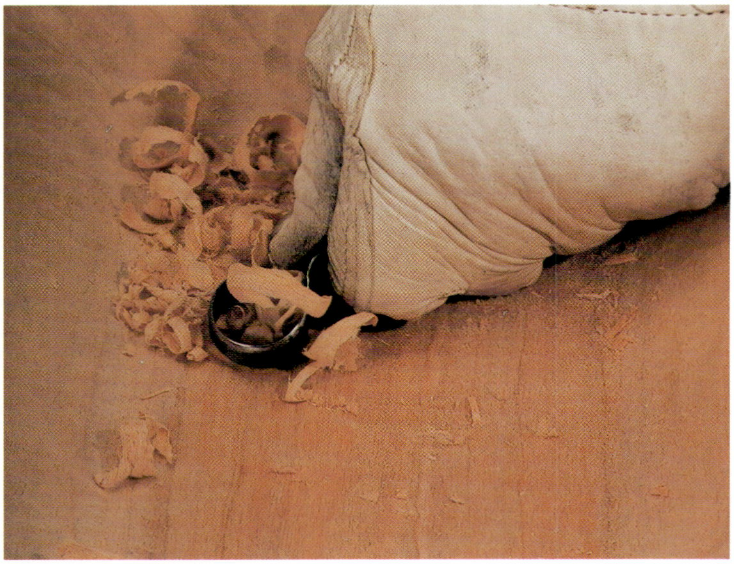

then down the length of the bottom.

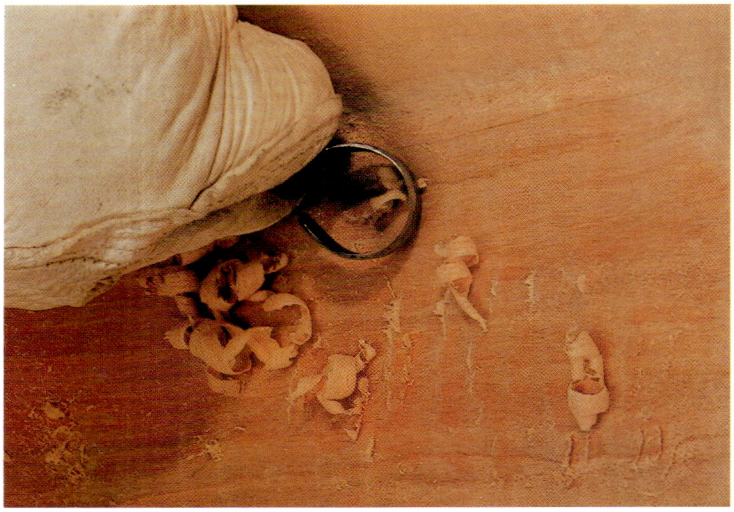

When you have achieved a flat, uniform surface, go with the grain for a final smoothing.

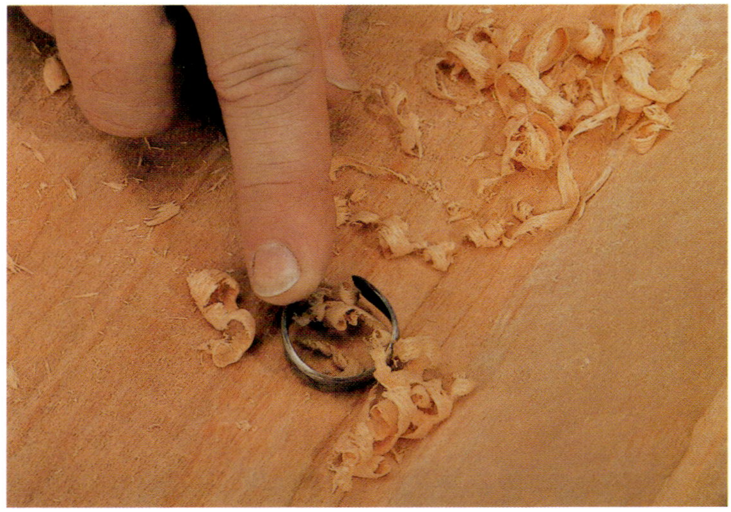

You need to go with the grain to avoid tearing. Of course the grain will shift as it has here, and you'll need to change directions several times to get it all.

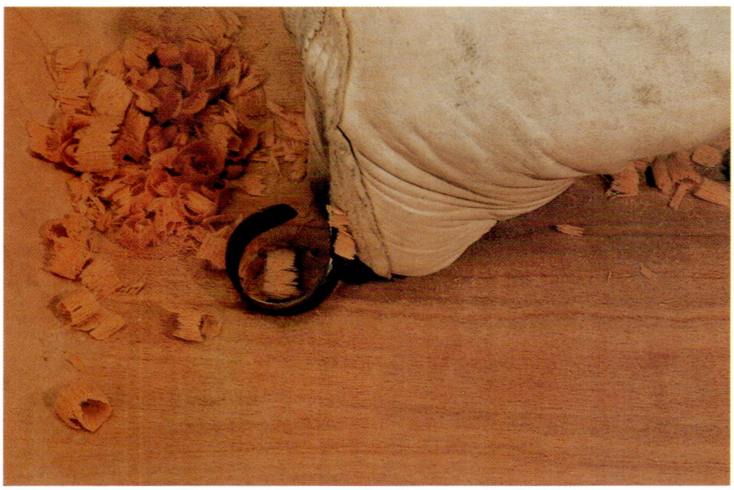

Go across the grain again if necessary.

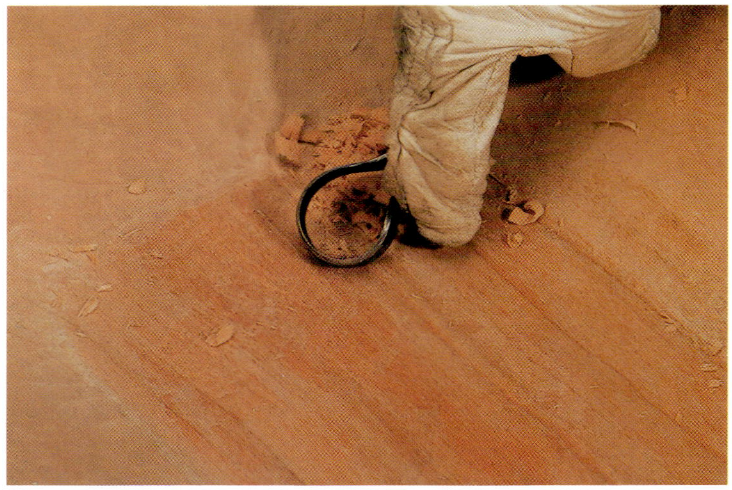

This last bit I clean up with an angled scrape.

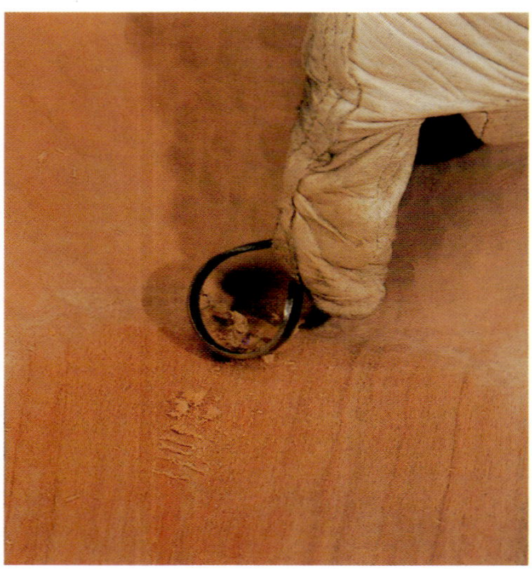

A final scrape with the grain...

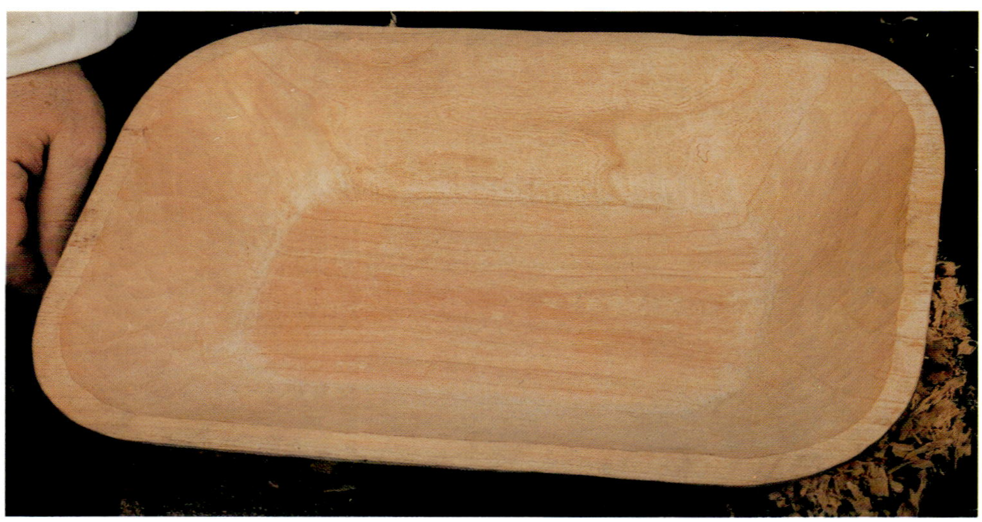

and it is ready for the finish.

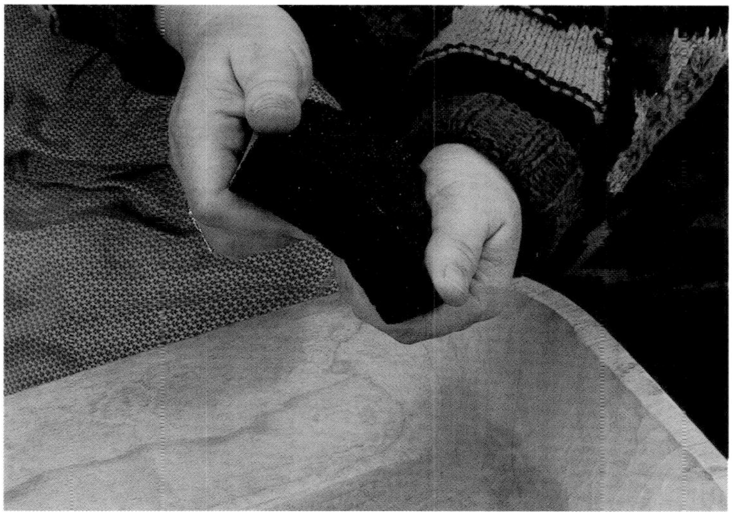

The adze raises the grain, so before oiling we need to close the wood grain. To do this I use a coarse/medium grit sanding sponge. This does not remove the tool marks, as a heavier sanding would do. On this bowl, with its fine grain, I will only use the medium grit. With heavier grains I would go over the piece once with the coarse grit, followed by the medium grit. You could use steel wool for this step.

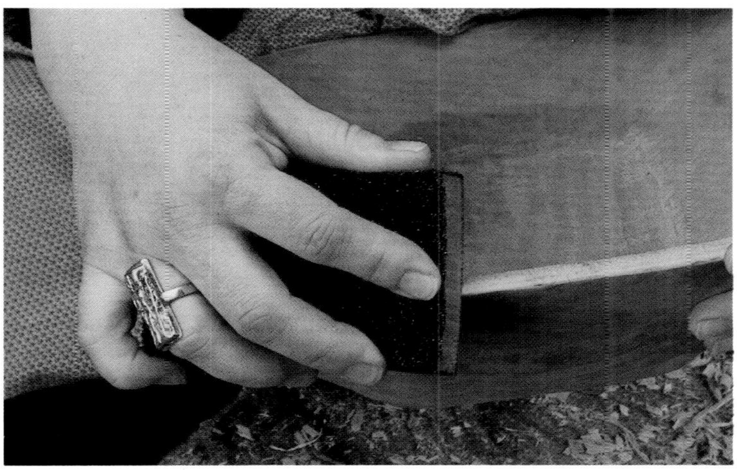

On the end grain you need to be a little firmer than on the side grain.

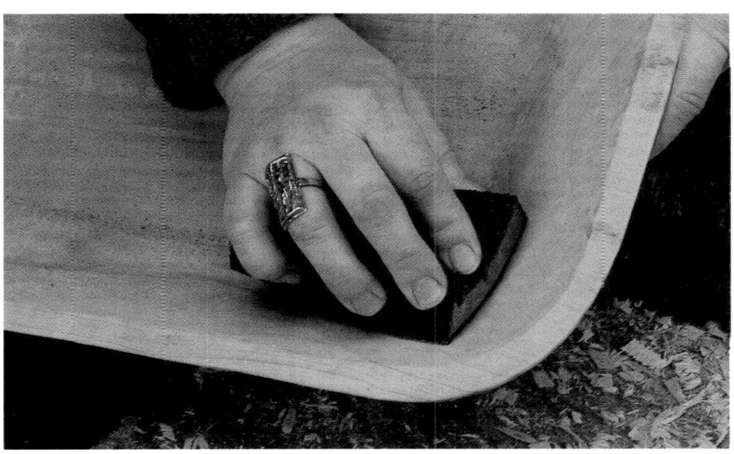

Moving to the inner wall of the bowl I usually start in the corner. This is where the flexibility of the sanding sponge is most appreciated.

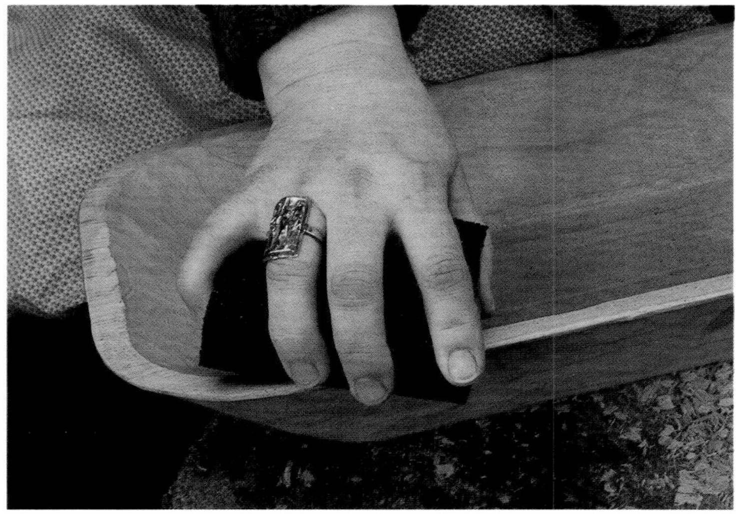

I always start on the rim. This is the part that nearly everyone first runs their hands over when they pick up a bowl, so I want to be sure it is pleasing to the touch. Begin with the outside edge going all the way around.

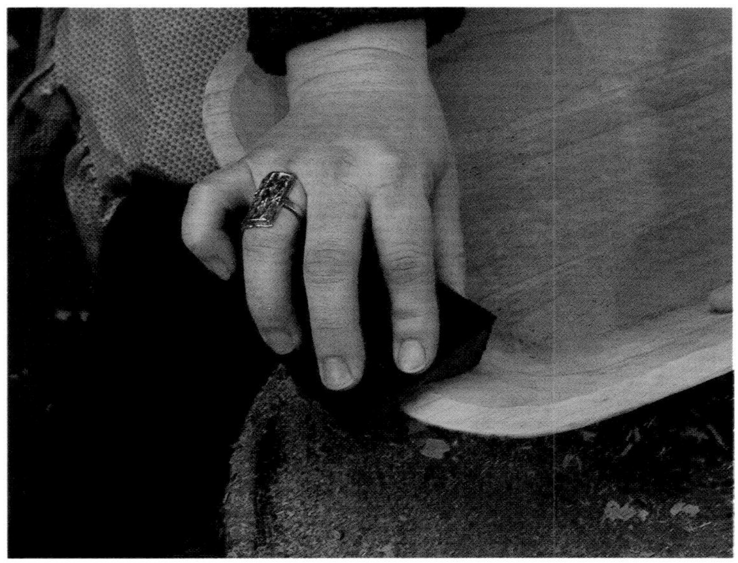

Next I do the top surface and inner edge of the rim.

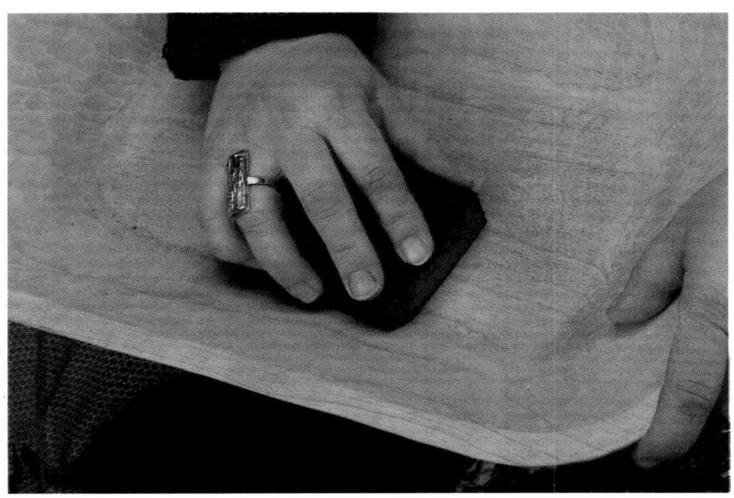

The bottom and corners finish the inside of the bowl.

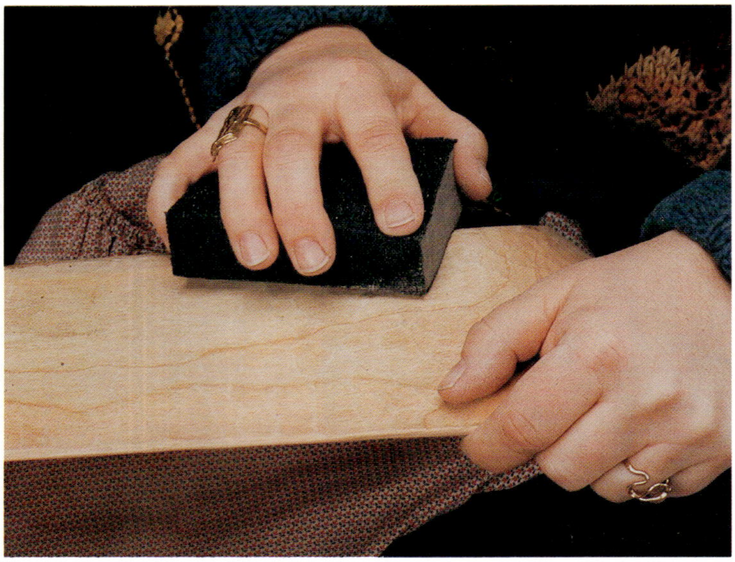

Work the outside of the bowl.

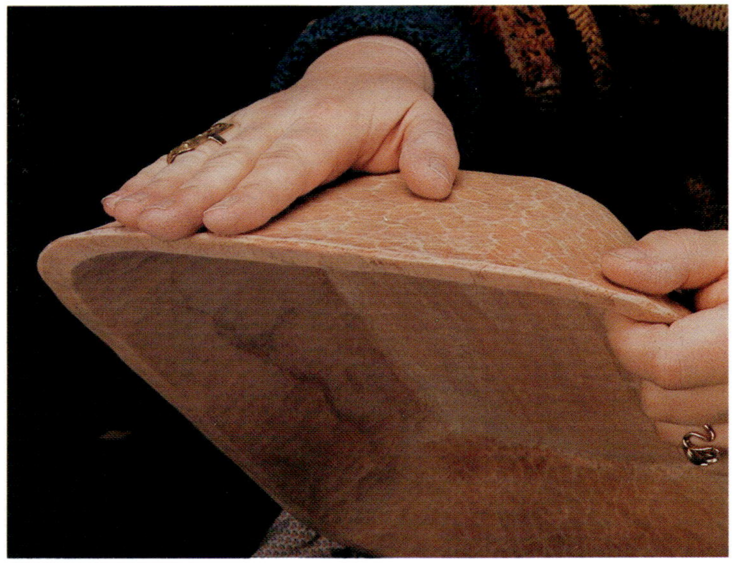

I go over the whole piece with my hand to check for smoothness.

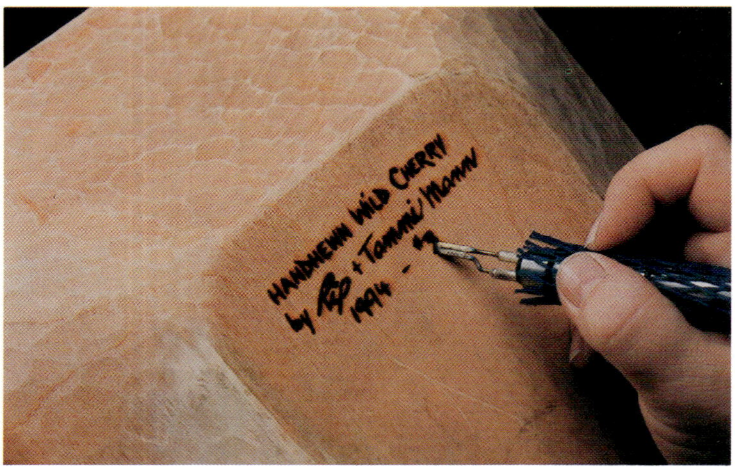

Before finishing you will want to sign your work. A few generations down the road, your great great grandchildren will look at the bowl and know who did the work and when. I use a woodburner for signing, but you could also carve the signature. Include your name, the date, and the type of wood.

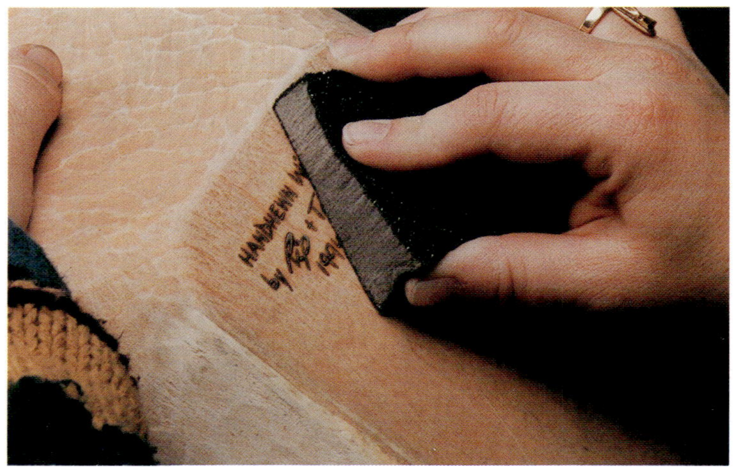

Go over the signature with the sanding block. We are now ready to begin the seasoning process.

The Seasoning Process

1. The oiled bowl sits in the bag in a cool dark space for 12 hours. Be sure it is fully supported on a flat surface.

2. At the end of 12 hours, take it out of the bag, allow the water that has beaded up to evaporate (approximately 20-30 minutes). Then liberally oil the wood again. Leave it out of the bag for 12 hours, checking every hour or so to see if it needs more oil, especially on the end grains. Reoil as needed during the day.

3. At the end of the twelve hours, oil liberally and place the bowl back in the bag for another twelve hours.

4. Repeat this process for the next three to five days. You will notice that the bowl requires less and less oiling during the twelve hours it is out of the bag. Denser woods will take longer than less dense woods.

5. At the end of the five days you will no longer need the bag. Leave the bowl in the same cool, dark place and check daily to see if it needs more oiling.

6. Leave the bowl in the cool dark place for approximately six weeks. Check periodically as it may need re-oiling. At the end of this time the bowl is seasoned. It is ready be used for food. Use often! We recommend reoiling every six months, and suggest Father's Day and Christmas as memory joggers. After using the bowl you may wipe it out with a hot sudsy cloth, rinsing and drying immediately. If especially messy, you can clean the bowl in the sink, but don't let it soak in the water. Rinse and dry immediately.

7. Don't leave the bowl on top of the refrigerator, in direct sunlight, or other places where heat or light may damage it.

8. Some woods, like walnut, feather up when exposed to moisture. Should this happen, it is helpful to go over the bowl with a light sanding.

The most important step in finishing the bowl is the seasoning. Because we work with green wood, if this is not done correctly the bowl will check and warp. Basically what we are doing is replacing the water in the wood with oil, either peanut or mineral oil, which are both non-toxic and do not go rancid. A couple of tablespoons of peanut oil is poured into the bowl.

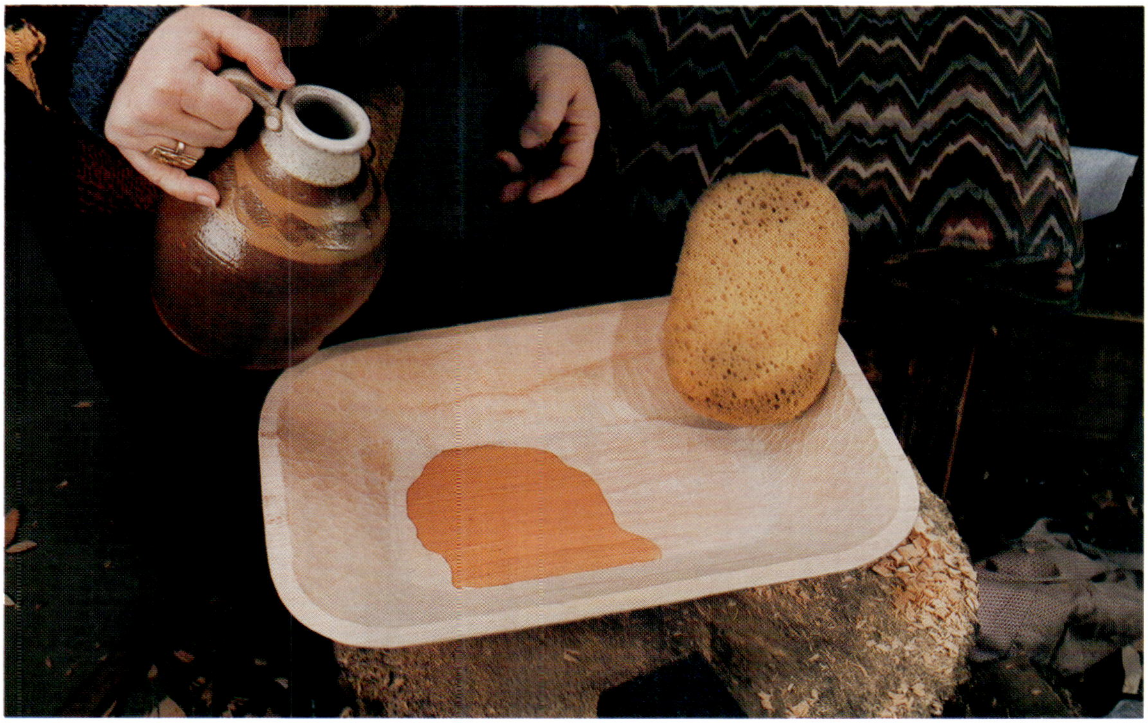

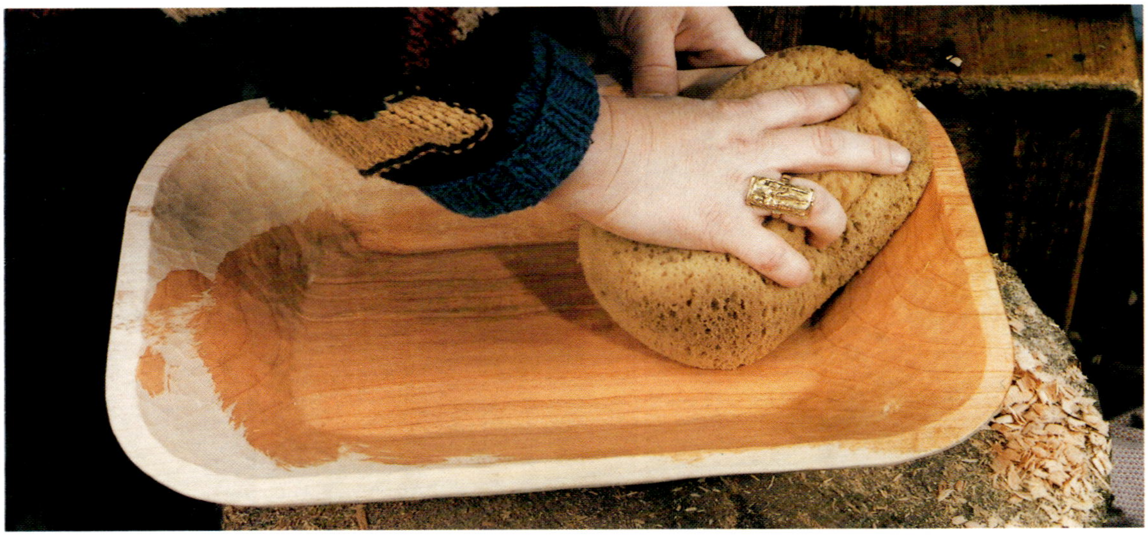

Cover all the surfaces inside and out with a liberal coating.

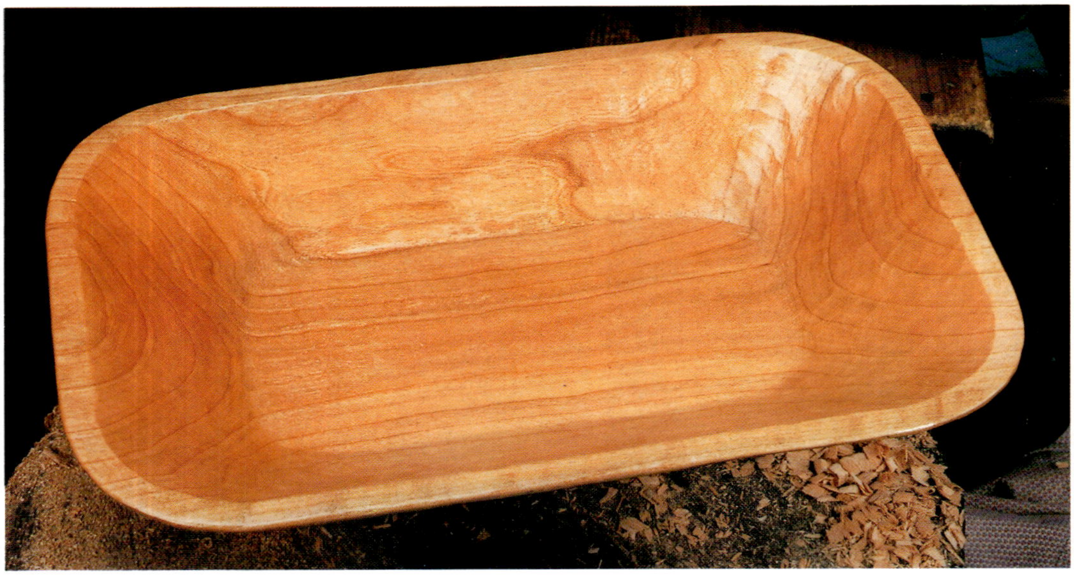

The first coat brings out the color of the wood.

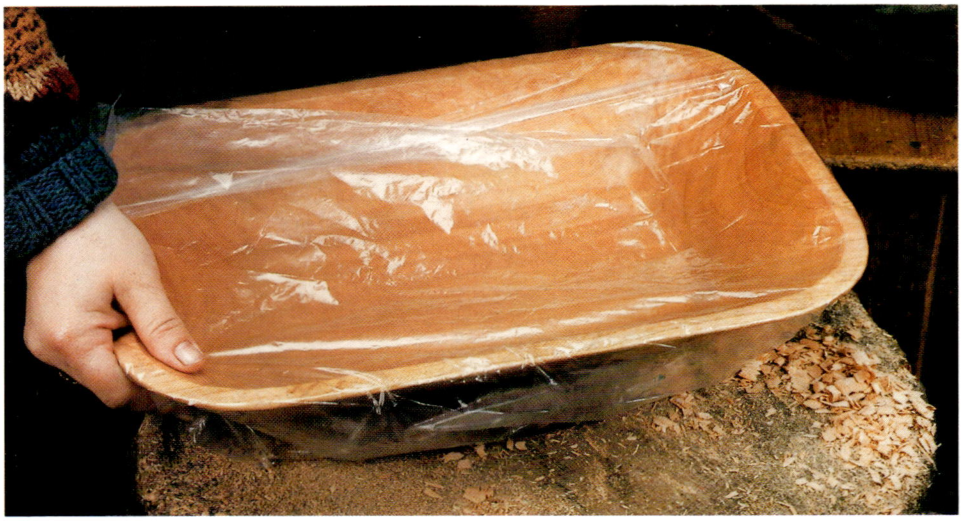

Place the bowl in a plastic bag, seal it with a twist-tie. The process involves moving bowl in and out of a plastic bowl. The bag slows down the drying process, allowing the slow replacement of the water with oil. In addition the bag allows the end and side grains to dry more equally.

The Gallery

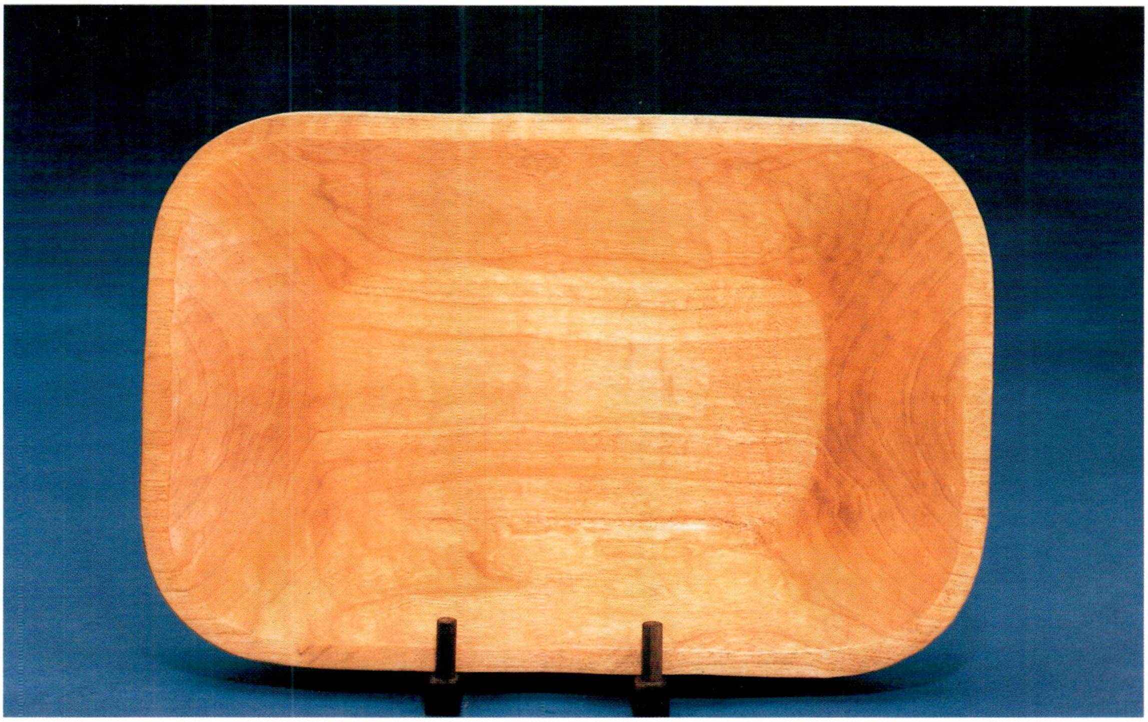

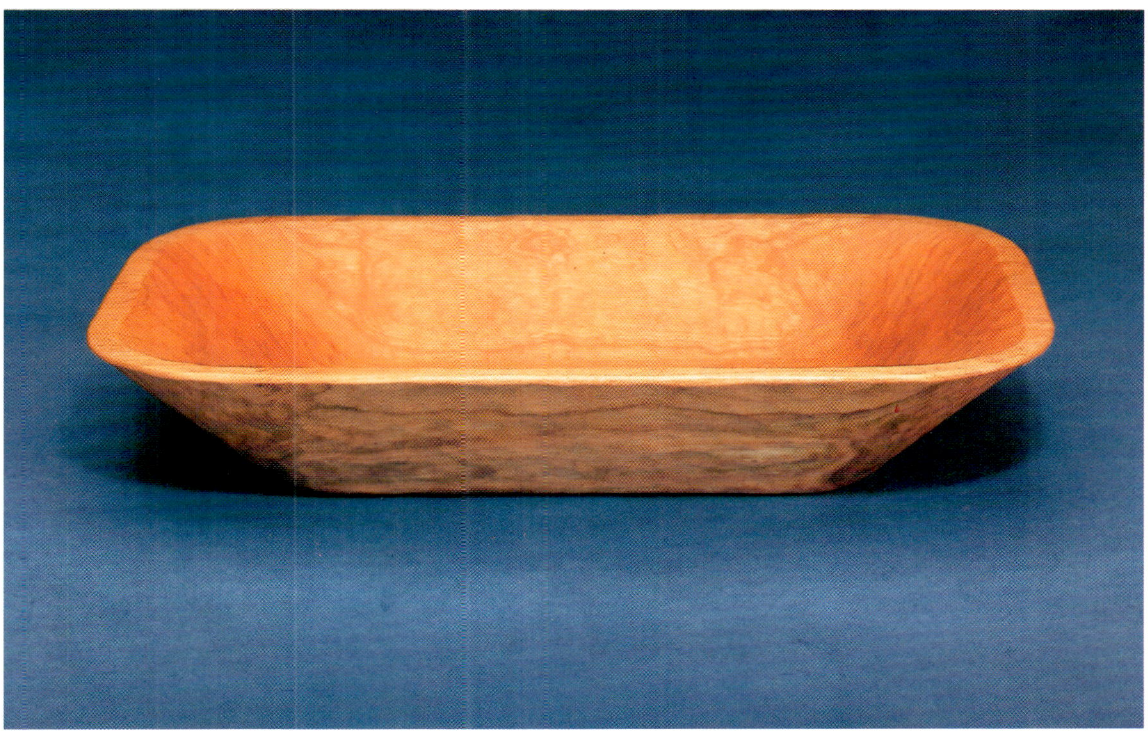

Appalachian Rectangle, Wild Cherry, 11" x 16" x 3"

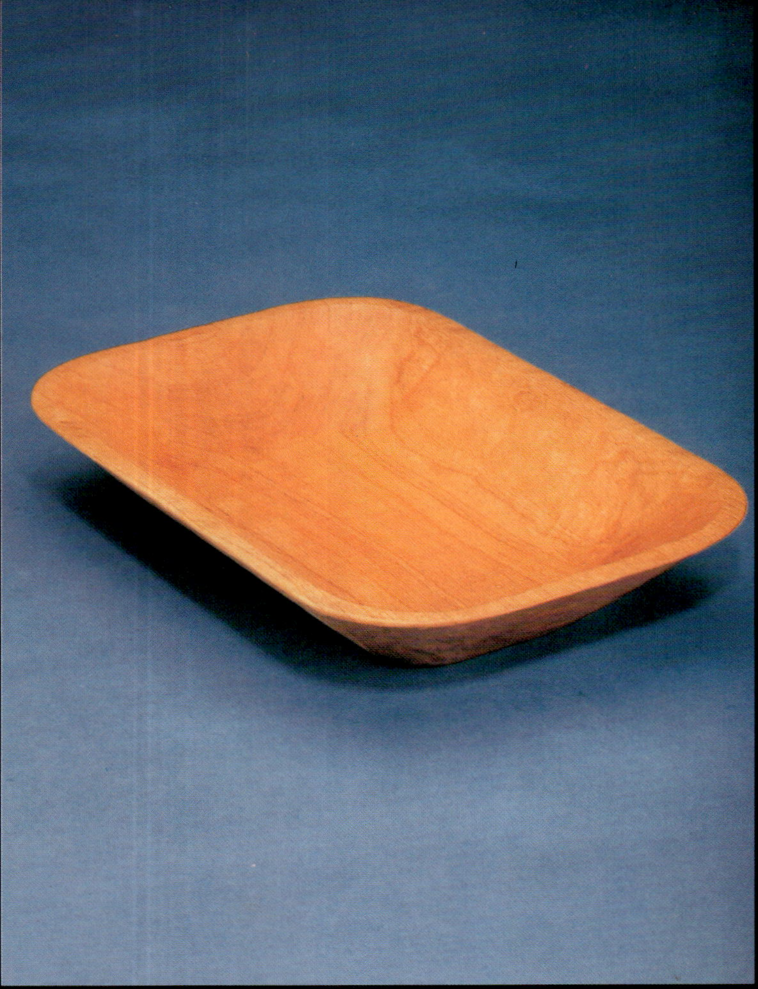

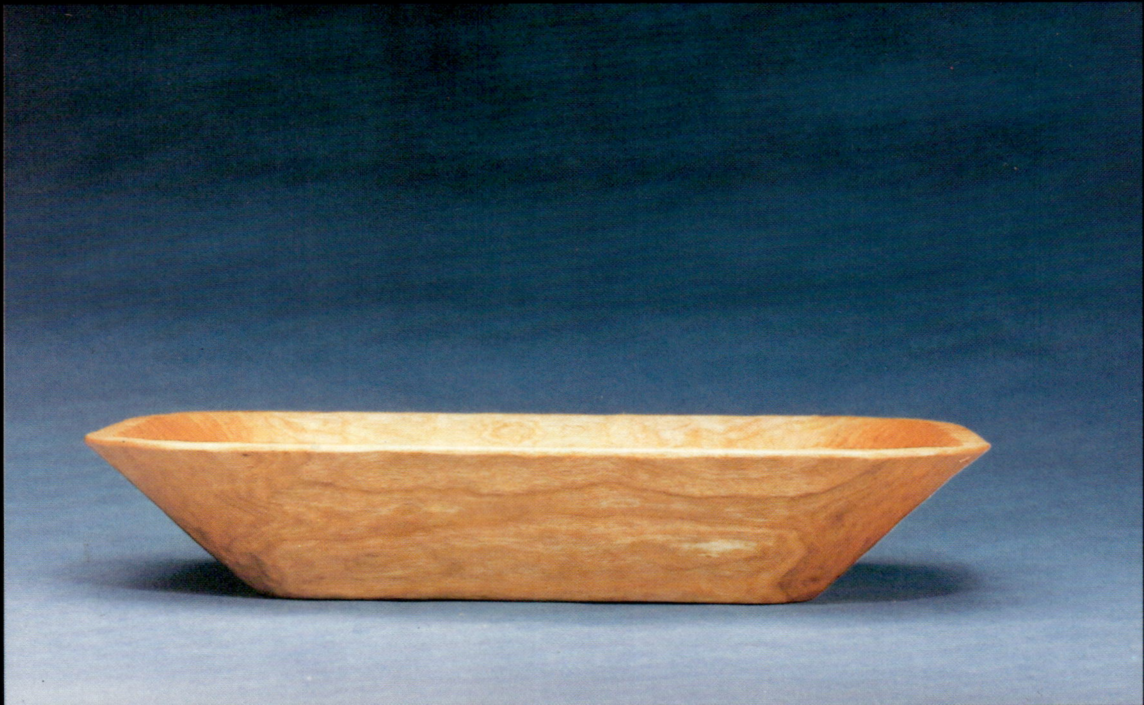

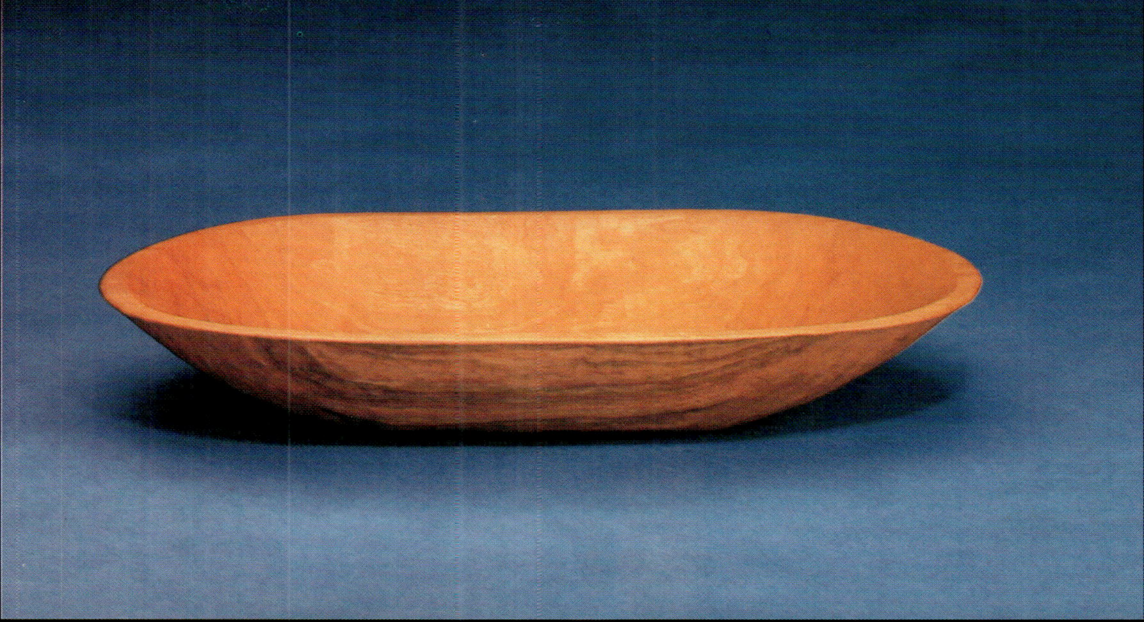

Dough Bowl, Wild Cherry, 14" x 22" x 4"

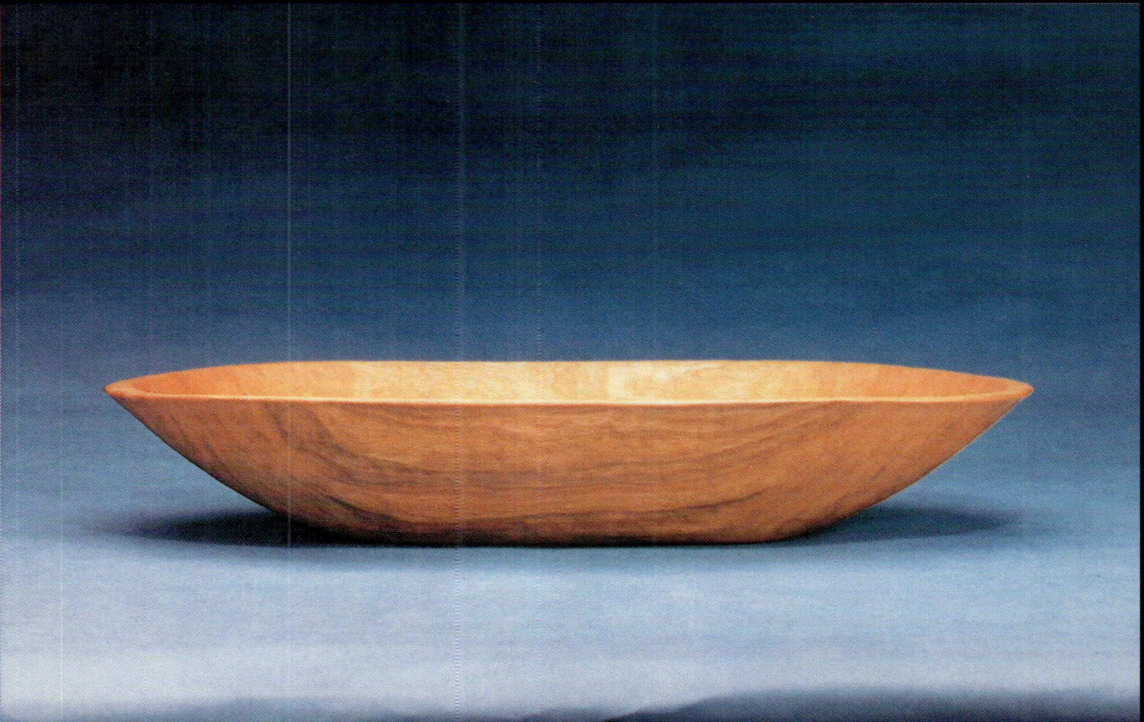

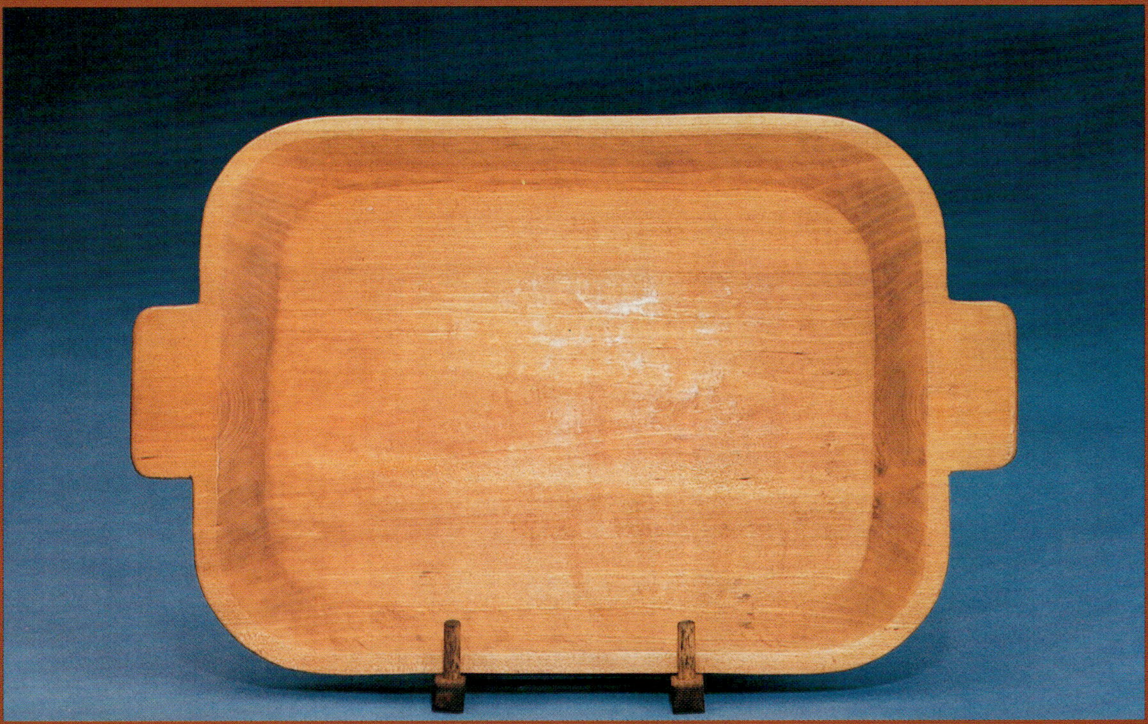

Serving Platter with Handles, Wild Cherry, 12" x 16" x 1 1/2"

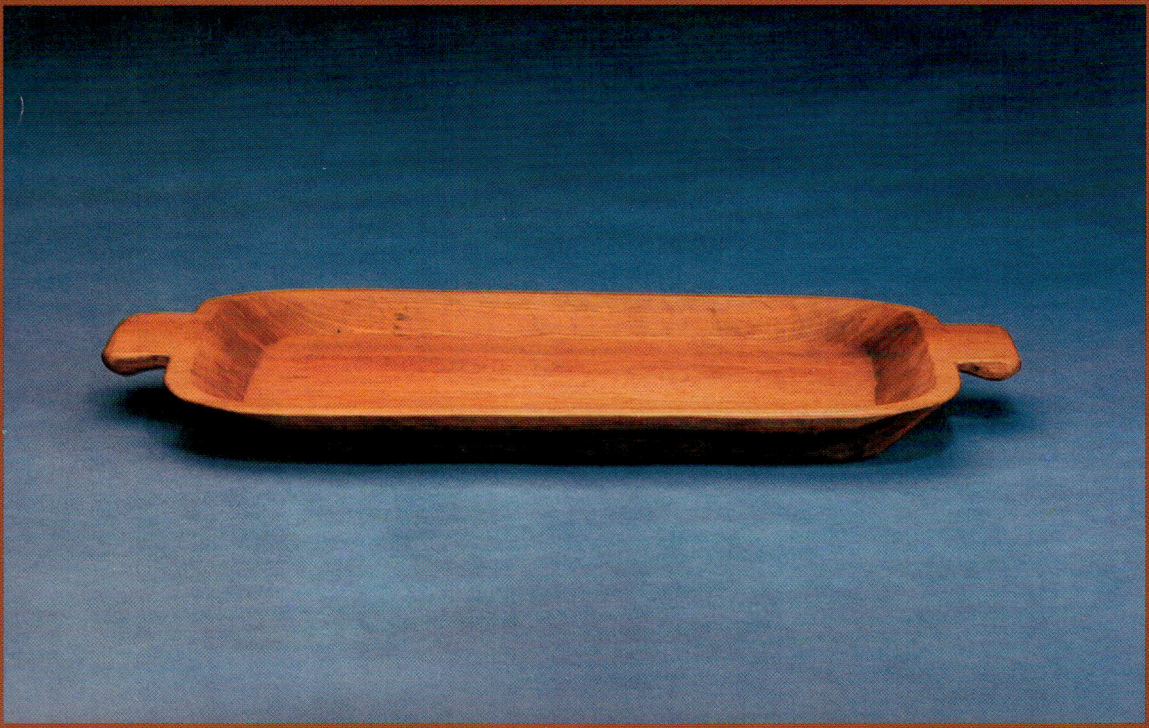